Mergers and Acquisitions in the Energy Sector

Corporate Finance and Governance

Herausgegeben von Dirk Schiereck

Band 14

PL ACADEMIC RESEARCH

Robert M. Fraunhoffer

Mergers and Acquisitions in the Energy Sector

The Impact of Synergy Disclosures on Shareholder Wealth and Operating Performance

PL ACADEMIC RESEARCH

Bibliographic Information published by the Deutsche Nationalbibliothek
The Deutsche Nationalbibliothek lists this publication in the Deutsche Nationalbibliografie; detailed bibliographic data is available in the internet at http://dnb.d-nb.de.

Zugl.: Darmstadt, Techn. Univ., Diss., 2013

Library of Congress Cataloging-in-Publication Data

Fraunhoffer, Robert M., 1986-
Mergers and acquisitions in the energy sector : the impact of synergy disclosures on shareholder wealth and operating performance / Robert M. Fraunhoffer.
pages cm. — (Corporate finance and governance, ISSN 1869-537X ; Band 14)
ISBN 978-3-631-64521-5
1. Energy industries—Mergers. 2. Corporations—Valuation. 3. Consolidation and merger of corporations. I. Title.
HD9502.A2F73 2013
333.79—dc23

2013025853

D 17
ISSN 1869-537X
ISBN 978-3-631-64521-5 (Print)
E-ISBN 978-3-653-03575-9 (E-Book)
DOI 10.3726/978-3-653-03575-9

PL Academic Research is an Imprint of Peter Lang GmbH.

Peter Lang – Frankfurt am Main · Bern · Bruxelles · New York · Oxford · Wien · Warszawa

www.peterlang.de

Preface

In kaum einer Branche wird bei der Ankündigung von Unternehmenszusammenschlüssen freiwillig so häufig über erwartete Synergien berichtet wie in der Energiewirtschaft. Das kann verschiedene Gründe haben und ist nicht nur spannend, weil gerade die europäische Energiewirtschaft durch den Ausbau von Erzeugungskapazitäten aus erneuerbaren Energien und dem teilweisen Ausstieg aus der Atomwirtschaft nach der Katastrophe in Fukushima dabei ist, sich neu zu strukturieren und dabei Unternehmenskäufe eine wichtige Rolle spielen.

Aus theoretischer Sicht sollte die freiwillige explizite Ankündigung von erwarteten Synergien keinen Einfluss auf den Börsenwert des Käuferunternehmens haben, wenn diese Information leicht abschätzbar und im Erfahrungswissen der Marktteilnehmer grundsätzlich verankert ist. Aber passiert das auch? Reagieren die Aktienkurse der Transaktionspartner bei Unternehmensakquisitionen unabhängig von der Bekanntgabe über erwartete Synergien? Entsprechende Untersuchungen zum Erfolg von M&A-Transaktionen unter Einbeziehung von freiwilligen Synergieankündigungen sind bis heute rar, der Kenntnisstand insbesondere zum Erfolg von Akquisitionen in der Energiewirtschaft ist ähnlich begrenzt und die Übertragbarkeit der Evidenz zu den Erfahrungen aus anderen Branchen ist mehr als fraglich. Eine umfassende Analyse der Erfolgsdeterminanten von M&A in dieser Branche und den Konsequenzen von Synergieankündigungen ist mir bis dato nicht bekannt.

Die vorliegende Arbeit nimmt sich dieser bedeutenden Forschungslücke mit viel Liebe zum Detail und höchster Sorgfalt an. Ihr primäres Ziel war es, anhand von Marktdaten den Erfolg von internationalen M&A-Transaktionen in der Energiewirtschaft zu ermitteln und wesentliche Determinanten dieses Erfolgs, insbesondere auch die Synergiepotentiale, zu bestimmen. So wird ein objektiver Kenntnisstand erreicht, auf dessen Basis sich fundierte Handlungsempfehlungen für die Industriepraxis ableiten lassen. Zudem finden sich damit auch für wirtschaftspolitische Entscheidungsträger wichtige Informationen zum Verständnis und zur Bewertung des sich weiter fortsetzenden Konzentrationsprozesses.

Herr Fraunhoffer kann die selbst gesetzten Ziele in seiner Dissertationsschrift bestens erfüllen. Die Arbeit enthält viele hoch interessante Resultate und ist so geschrieben, dass es den Lesern große Freude machen wird, sie Seite für Seite bis zum Ende zu studieren. Ich wünsche der Arbeit eine weite Verbreitung.

Professor Dr. Dirk Schiereck

Acknowledgements

The dissertation process would not have been possible without the support, advice, and help of many people. I would like to thank those people for their continuous effort as they enabled me to complete my dissertation in a highly focused and successful manner.

Firstly, my deepest gratitude goes to Professor Dirk Schiereck. Is was his support, his numerous recommendation, and critics that made me succeed with my dissertation. His highly valuable insights during my two years at the chair of corporate finance at the Darmstadt University of Technology were accompanied by his personal touch as a doctoral father which made the time highly enjoyable as well as a great personal experience.

Secondly, the enjoyable past years were also driven by the wonderful colleagues and supporting environment that accompanied me. A special thanks goes Daniel Maul with whom I shared an office. Daniel was not only always available for discussions on various topics of econometric analysis and scientific interpretations, but also provided me with thoughtful insights and recommendations that helped me in completing my dissertation. Moreover, I am as well grateful to Christian Babl, Anit Deb, Christian Happ, Steffen Meinshausen, and Julian Trillig who constantly showed a great support and interest in my academic work.

Thirdly, the dissertation would not have been possible without the encouragement of my family. My parents, Brigitte and Mathias, supported me again for which I am highly thankful. In addition, I am grateful to my sister Katharina who was always very helpful in supporting my objective. Finally, I would like to say thank you to Julia for her continuous support and moreover her encouragement which served me as a great source of motivation.

Robert M. Fraunhoffer

Content Overview

Table of Contents

List of Figures

List of Tables

List of Abbreviations

AR	Abnormal Return
BHAR	Buy-and-Abnormal-Return
CEO	Chief Executive Officer
CAR	Cumulative Abnormal Return
CAAR	Cumulative Average Abnormal Return
DEA	Data Envelop Analysis
EU	European
E(R)	Expected Return
FF3F	Fama-French-3-Factor-Model
GARCH	Generalized Autoregressive Conditional Heteroscedasticity
HICP	Harmonized Indices of Consumer Prices
HML	High Minus Low
kWh	Kilowatt-Hour
M&A	Mergers and Acquisitions
OCF	Operating Cash Flow
OLS	Ordinary Least Square
PV	Present Value
SMB	Small minus Big
RI	Total Return Index
ROA	Return on Assets
ROS	Return on Sales
US	United States
VIF	Variance Inflation Factors
WLS	Weighted Least Square

1 Introduction

Developments in the United States (US) and European (EU) energy markets in recent decades have, to a great degree, been driven by regulatory changes. Among the multiple government-enforced efforts, market deregulation was *the* fundamental initiative which lead to challenges, as well as opportunities, for energy firms. The following dissertation centers around these consolatory efforts and investigates the potential to create firm value via mergers and acquisitions (M&A) Firm value in the context of public firms is described as the creation of shareholder value. In this regard, two considerations are essential: not only does the M&A have to be value increasing from an internal, accounting-related firm perspective, but also the market has to perceive and subsequently price this value increase. Therefore, it might be beneficial to disclose additional information on merger gains besides the mandatory reporting schemes in order to convey rational market participants of a subsequent value creation. The gains that accumulate during M&A are mostly derived from synergy effects. The share price and subsequent value consequences of such voluntary synergy disclosures are the core focus of the present dissertation.

The appreciation of share prices due to M&A is dependent on the market's valuation as firm outsiders (i.e. the capital market) value firms' internal effects. This interdependency proposes certain challenges to managers; among these, information processing from the inside to the outside stands out. Thus, management has to ensure that share prices fully reflect the outstanding information, which suggests that managers should not only disclose sufficient and relevant information, but also do this in an efficient manner so as to enable capital markets to value it. In particular, within an unstable environment such as the energy sector, efficient information processing can help to decrease uncertainties for investors. In this regard, it may be beneficial if voluntary disclosures are released to the market in addition to mandatory reporting schemes. Mandatory reporting in general encompasses financial statements, footnotes, management discussions and analyses, as well as regulatory filings (Healy & Palepu, 2001). Such regulatory filings include, for instance, the pure announcements of M&A. Voluntary reports cover any additional and more detailed mandatory filing, and can either be reported along with a press release or through presentations or conference calls.

The necessity of additional information disclosures emerges as markets and investors are seldom fully aware of the management's objective; hence, the knowledge base of the outside capital markets and the inside firm is unequal, which leads to information asymmetries. These may be substantial or minor depending on the respective subject, yet their pure existence hampers the capital market's ability to correctly value a novel firm, which is consequently likely to result in a false value perception. Myers and Majluf (1984) conducted one of the first studies that shows such an effect. Their research reveals that the reduction of information asymmetries leads to a reduction of external financing costs. Hence, the market appreciates a more certain corporate environment. Comparable findings are provided by Merton (1987), who shows that the capital market's demand for premiums is directly related to the degree of information asymmetries. Other examples of beneficial developments from decreased information asymmetry stem from voluntary earnings forecasts. Baginski, Hassell, and Waymire (1994) find that positive voluntary earnings announcements are associated with an abnormal positive market valuation. Healy, Hutton, and Palepu (1999) further show that regardless of the type of voluntary disclosure – whether it be an earnings forecast, a conference call or any other – stock prices generally increase; thus, the market appreciates any attempt to decrease information asymmetries in routine situations. Gelb and Zarowin (2002) confirm this notion. A comprehensive overview of voluntary disclosure effects can be found in Healy and Palepu (2001).

Along with efficient information processing, a necessary precondition for positive capital market valuations is the underlying value of the M&A. The objective of these can be classified into neoclassical and behavioral theories (Bernile & Bauguess, 2010). Behavioral (or, synonymously, agency) theory proposes that M&A represent an expression of agency conflicts between firm insiders and investor outsiders. Thus, managers might pursue transactions where the objective is not to maximize the firm value, but to gain diverging personal benefits, which subsequently leads to firm value destruction. These personal benefits may relate to managers seeking to build an empire (Jensen, 1986), or increase their compensation (i.e. salary, bonus, or stock option) (Mueller, 1969); alternatively, they may be overconfident about actually realizing expected gains (Roll, 1986). On the other hand, neoclassic theory argues, according to Harford (2005), that M&A emerge in response to industry shocks, and hence are a reaction to external factors. These shocks encompass technological, industry, or regulatory changes, while the subsequent M&A is intended to overcome the emerging challenges. Therefore, the M&A objective is to increase the overall firm efficiency, thus leading to value gains via synergies.

As Berkovitch and Narayanan (1993) show, these competing theories are seldom mutually exclusive; thus, most mergers encompass a variety of objectives. Nevertheless, their research also demonstrates that emerging efficiency gains, i.e. synergies, are *the* primary merger objective as these create true firm value. Thus, M&A are subject to the creation of firm value on an aggregated firm level.

As the discussion above shows, voluntary disclosures enable a more effective capital market valuation; thus, their role for M&A shall be addressed in the following. It is particularly noteworthy that the communication strategy during takeovers differs significantly from routine situations. Standard merger announcements solely inform the market about the merger itself and the transaction premium, but not about the merger motive or the implied wealth gains. Therefore, information asymmetries are unusually high, which is reflected in negative announcement returns for the acquiring firm (Moeller, Schlingemann, & Stulz, 2007; Officer, Poulsen, & Stegemoller, 2009). Yet, according to Luo (2005), the market's evaluation, particularly a positive evaluation, of the merger is an essential aspect for successful merger completion. Thus, the attempt to decrease information asymmetries during M&A and thereby inform the market of subsequent value creation should lead to beneficial developments for the bidding entity and the transactions as a whole. Such additive information measures are voluntary disclosures, as stated above. In this light, Kimbrought and Louis (2011) show that voluntary conference calls during M&A announcements lead to positive returns. Their evidence suggests that these calls are primarily utilized to convey genuine – and therefore private – information in a credible manner, so as to reduce information asymmetries.

Besides the emphasis on forward-looking details, the management needs to disclose information which can be rationally priced as part of a firm's value. With respect to emerging synergy effects during M&A, this demands the provision of a quantitative forecast. Thereby, it is necessary not only for information asymmetries to be reduced, for example via conference calls, but also for value be communicated. Thus, for rational market participants, a subsequent increase in returns should result. However, the synergy potential and magnitude, as well as the disclosure rate, differs significantly between various industries. In a cross-sectional study, Bernile and Bauguess (2010) show that disclosure rates vary from 10% for producers to 42% for utilities. Hence, in the present study the focus is laid on a concrete industry, i.e. the energy sector, but on a global basis.

This focus on the energy sector is driven by a variety of reasons. First, this industry has experienced a substantial wave of mergers in recent years. As Figure 1.1 illustrates, the number of M&A transactions has steadily increased during recent decades; the number of transactions has doubled from almost 2,000 in 1994 to more than 4,000 in 2011. This volume can be expected to remain high as traditional energy sources increasingly combine their operations with renewable ones. This transition is partially government-enforced, such as by the EU initiated "20/20/20" objective (i.e. improve energy efficiency by 20%, increase the renewable energy share to 20%, and reduce greenhouse gas emission by 20%), yet is also strategically driven (Stankeviciute & Criqui, 2008).

Figure 1.1: M&A in the global energy sector

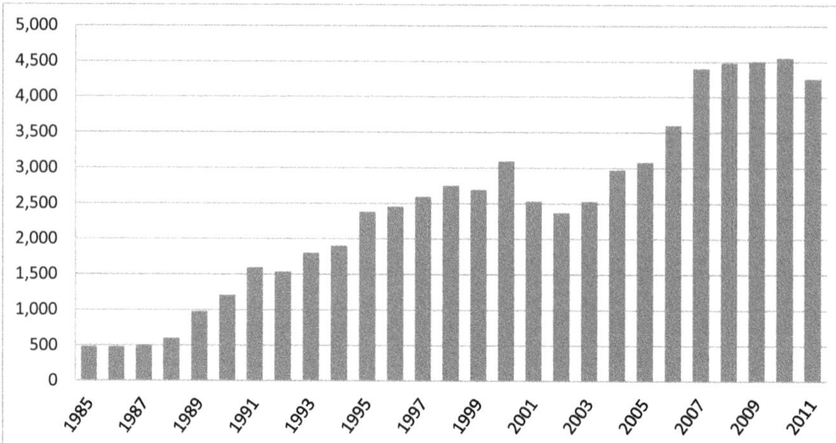

Figure 1.1 shows the number of all transactions in the global energy market from 1985 to 2011. Data is provided by the IMAA (2012).

Secondly, the US and European energy markets have undergone substantial deregulation. This deregulation has, to a great degree, enabled and also fostered the increased amount of mergers. In Europe, Directive 96/92/EC, issued by the European Council in 1996, initiated this development while acknowledging that certain market segments should be liberalized. Therefore, networks are unbundled and monopolies abolished while energy firms are allowed to merge across the value chain. Continuous deregulatory initiatives further promote this development. One of the most recent initiatives is the third energy package from 2007; its objective is to support the transition towards product bundles, i.e. combine various energy forms and thereby serve existing customers with multiple energy sources. In the US, similar deregulatory efforts emerged with the introduction of the Energy Policy Act in 1992. Owing to these developments, the energy sector represents a highly interesting research field, as merger opportunities that have been unexplored to date can be realized.

Lastly, the energy sector is also characterized by a naturally high degree of synergy potential, and therefore shows an increased disclosure rate. The following research reveals that 45% of all energy-related M&A transactions disclose synergy forecasts. As the disclosure effects are investigated, a sufficient dataset is provided which not only infers robust implications, but also extracts meaningful firm-level recommendations.

On the whole, this dissertation combines the aspects elaborated above and analyzes the value-creating potential which stems from voluntarily disclosing synergy forecasts during M&A, and their post-merger realization in the US, as well as the EU, energy markets. Thereby, within this dissertation, three consecutive studies take different perspectives to conduct a comprehensive analysis. Initially, a practitioner's perspective is assessed, which is in a following investigation extended by the capital market's assessment on a short and long term basis. Lastly, the actual synergy realization is analyzed, and these forecasts are tested for their credibility in an accounting context. The respective research titles are:

- Consolidation and Synergies of the Energy Sector: New Insight from Top Managers
- Synergy Communication Effects in European Energy Markets
- The Interaction of Synergy Disclosures and Post-Merger Operating Performance

The first study (Chapter 2) investigates the synergy potential of the energy sector. Thereby, the focus is specifically laid on the German energy market, as Germany is not only the largest electricity market in the EU, but also has not been researched thoroughly in the past, as the majority of prior synergy-related research has been completed for the US economy. Hence, this study will advance the field in a novel way while a new methodical component is introduced to synergy research, i.e. expert interviews. The subsequent objective of this chapter is twofold: initially, a literature review on synergies in the energy market is conducted to provide a comprehensive overview. This empirical, objective evidence is then supplemented by the subjective perspective of practitioners. Therefore, three open-ended expert interviews are conduted with the three leading, publicly traded energy providers in Germany. This not only provides new evidence to contribute to the ongoing discussion, but also a novel perspective, as it is not *ex post*, realized synergies that are reported (as in empirical studies) but *ex ante*, expected ones. This approach hence makes it possible to analyze aspects that have remained unexplored to date.

Overall, the evidence suggests that financial synergies are of minor importance. Scale and scope economies are especially relevant for smaller utility firms, while these synergies particularily emerge in distribution and generation functions. Operational synergies, which are perceived to be the second-most relevant synergy source, also emerge in distribution functions, yet also in the retail segment.

The second study (Chapter 3) investigates whether voluntary synergy disclosures are an efficient instrument to decrease information asymmetries. If so, an increase would be seen in the short-term announcement returns relative to those firms not disclosing synergy forecasts. In order to detect these, the short-term event study methodology is utilized, while the sample is extended from Germany to the whole European energy market. To further capture disclosed synergies, M&A announcements are hand-collected, and each firm's expected amount of synergies is individually discounted to the time of the announcement to ensure their comparability. In addition, this chapter tests whether increased short-term returns remain robust over a long-term period. Therefore, the Fama-French-3-Factor (FF3F) model is utilized, as well as the buy-and-hold-abnormal-return (BHAR) methodology. Hence, overall, the study extends the literature on voluntary disclosure generally (as the first study that tests voluntary synergy disclosure effects in the European market), and the synergies at the merger announcement specifically.

The results show that energy sector synergies significantly differ from cross-industry ones, which again underlines the neccessity to conduct sector-specific analyses. In addition, the expected postive effects of voluntary synergy disclosures can be confirmed, yet solely for the bidding firm: acquirers experience increased short-term returns while disclosing synergy forecasts; targets, on the other hand, do not. Nevertheless, the combined entity does lead to value creation, and hence synergy disclosures efficiently reduce information asymmetries. Furthermore, the results show that these firms are able to sustain their abnormal performance in the long-term. Thus, the market does not overreact in the short-term, and management projections are apparently provided in a reasonable manner.

The third study (Chapter 4) picks up on the work conducted in the prior chapter and extends the research towards the US market. Initially, the capital market's response to voluntary synergy disclosures is investigated in order to confirm the effects found for the EU in another environment. The subsequent objective of this chapter is to investigate the operational reasonabilty of these synergy forecasts. The prior chapter considers a capital market perspective on which to judge this; however, emerging long-term positive or negative returns can solely provide an indication for the actual realization of synergies. Stock market returns are driven by a variety of factors, and their abnormal increase does not necessarily imply the actual realization of disclosed synergy projections. Therefore, an operationally driven analysis is conducted, which excludes the market's valuation while utilizing two methodologies, i.e. the change and intercept model. Thereby, M&A are distinguished according to their motive to realize efficiency gains, and it is subsequently expected that these transactions alone experience a performance increase. Further investigation is conducted with respect to those characterisitics that the management believes drive the realization of synergies, and those factors which actually foster a performance increase.

The evidence confirms that voluntary synergy disclosures in the US energy sector also lead to inceased returns at the announcment of the M&A. The operating performance study shows that the M&A motive distinctively drives the post-merger performance. Those transactions which did not expect synergies not only underperform size- and performance-matched non-merging firms, but also those firms that did disclose synergies. These, in contrast, show no significant difference relative to comparable firms; hence, no performance loss manifests. Thus, M&A can lead to value generation if the transaction and following integration is structured accordingly. In addition, the results show that the value-drivers identified by management do not necessarily have an influence in reality – for instance, the relative target and bidder size. This parameter reveals a strong influence on the management synergy projections, yet no effect on the operational post-merger performance.

Overall, the dissertation leads to three main implications for practitioners. Firstly, synergies in the energy sector can not only be realized during M&A, but are also essential. However, their relevance stems from two considerations. On the one hand, a value appreciation may particularly result from operational, as well as scale and scope, synergies. On the other hand, practitioners indicate that synergies are also utilized as a political instrument apart from value generation. Hence, synergy disclosures also serve as a support function, as these forecasts further increase the likelihood of the merger realization.

Secondly, synergy disclosures are an efficient instrument by which to decrease existing information asymmetries during M&A in the energy sector. Regardless of the market environment, M&A with an associated disclosure lead to increased announcement returns for the bidding firm. Therefore, the capital market attributes synergy realization benefits, and thus value appreciation, to the acquiring entity.

Thirdly, synergy-disclosing transactions are able to meet expectations. On a long-term perspective these firms are able to outperform their counterparts from an operational, accounting-based perspective, and therefore do not destroy value. Shareholders recognize this, and thus the capital market perspective confirms the operational study, as long-term returns are also significantly higher for synergy-disclosing M&A.

These results further imply for academics that empirical studies should not consider all M&A transactions to be equal. The merger motive apparently has a distinctive influence on the market's merger perceptions. Thus, if empirical research does not control for this, the results are likely to produce a false perception and ultimately result in misleading conclusions.

2 Consolidation and Synergies of the Energy Sector: New Insight from Top Managers[1]

2.1 Introduction

In 1879, a new industry was born: the energy industry. This development was initiated by the invention of the light bulb by Thomas Edison, and thus the energy industry was originally labeled the illumination industry. In its early decades the market was highly fragmented, with multiple providers for similar services, as well as inefficient firm structures. This, as well as other factors, led the market to consolidate, in particular through an increased rate of M&A. Later developments which fostered more efficient market structures included changes in government-issued regulations that were intended to also establish clear market structures (Froelich & McLagan, 2008). More recent market adjustments are targeted at deregulating the once highly regulated energy markets, which suggests that increased uncertainty and an unpredictable market will follow for utility firms. The primary objectives of the deregulation efforts are the facilitation of competition while allowing energy providers to move into all links of the energy value chain, as well as to merge or acquire energy firms that operate with different energy sources. Once the US began their efforts to deregulate (Hess, 2010), the EU followed in 1996; this led here to an 80% increase in the energy-related M&A rate in 2004/2005, while the global energy-related M&A rate rose by just 30% (Datamonitor, 2005). Obviously the European market, whose key players are in Germany, France, and Italy, appreciated the deregulatory activities as a means to grow.

A critical success factor in M&A transactions is the bidder price, which should not exceed the expected benefits. In the light of this aspect, the transaction premium has been identified as a crucial explanatory factor for revaluation of the bidders and targets (Moeller, Schlingemann, & Stulz, 2004). M&A failure in this context is defined as the inability to generate shareholder value, while it has been suggested that the shareholder wealth of the bidding party is particularly affected. In addition, a quadratic relationship is proposed between premium and bidder returns. That is, the premium has a positive influence on the shareholder value, up to a certain level at which it becomes negative (Diaz Diaz, Sanfilippo Azofra, & Lopez Gutierrez, 2009). Thus, the destruction of value

---

1 I am grateful to Prof. Dr. Dirk Schiereck for helpful insights and comments. An abbreviated version of this chapter has been published in Problems and Perspectives in Management, Volume 10, Issue 2, pp. 68-79.

does not result from the management's failure to capture the target firm's value, but rather from the inability to not only capture, but also evaluate the potential synergies. By missing synergies completely, or by overstating their magnitude and thereby paying an excessive premium, an M&A transaction is set to fail (Kode, Ford, & Sutherland, 2003). Madura and Ngo (2008) find evidence that premiums, and thus also expected synergies, are closely associated within industries, and highly depend on the synergies of prior transactions. Moreover, Hambrick and Hayward (1997) conclude that the premium is largely reflected by the bidder's firm performance, the media presence of the chief executive officer (CEO), as well as a measure of the CEO's self-importance. These findings highlight two critical incidences: being able to capture the actual synergy size is of utmost importance; however, firms apparently do not consider an operational approach for capturing synergies, and thus are likely to overstate them.

Each industry, with its unique characteristics, has unique sources of synergies which need to be explored individually. Such a synergy investigation has been conducted, for instance, within the Taiwanese banking industry (Ting-Kun, 2010), the Indian pharmaceutical industry (Chaturvedi, Chataway, & Wield, 2007), and the real estate industry (Campbell, 2002). Studies on synergies within the energy industry have either focused on deregulation effects in the US (Sueyoshi, Goto, & Shang, 2009) or on combinations of various energy sources (Fraquelli, Piacenza, & Vannoni, 2004). Overall, however, a limited amount of empirical research exists, and this has been partially outdated, with studies ranging back to the 1980s. The following examination will initially summarize these energy-related synergy studies, and thereupon extend this field of research with two additional aspects: first, an additional data collection method is introducted for energy-related synergy research through interviews with top managers in the energy industry; second, a new geographical focus is considered. All previous studies consider an empirical, objective research perspective; thus, based on quantitative capital market oriented data, M&A motives and synergy drivers are inferred. Following this deductive, indirect approach, vital insight is potentially lost, which most likely contributes to the inconclusive evidence found regarding synergies in the energy sector. As such, Filippini (1996) along with Sueyoshi and Goto (2011), for instance, conclude a rather small to negligible importance of energy-related synergies, while Yatchew (2000), as well as Piacenza and Vannoni (2009) attribute these to obtaining high relevance. By considering a direct, interview-driven research design, the present study not only contributes new evidence to the ongoing discussion, but also provides a novel perspective, as it does not report *ex post*, realized synergies without concrete knowledge of their sources, but rather provides *ex ante*, expected ones. This approach makes it possible to analyze thus-far unexplored aspects, which additionally serve as both supporting and contrasting elements regarding the existing evidence. In addition, Germany, as the economically most powerful country in the EU, which moreover contains the largest electricity market in Europe (Pereira da Silva & Soares, 2008), has been neglected by prior research, and hence shall be analyzed here. Thus, as the majority of the existing literature is concerned with the US market, new evidence regarding another highly relevant marketplace contributes to the diversity of this field of research, and further advances the current discussion towards a more coherent perception of synergies.

On the whole, the examination of this chapter extends the research upon synergies in general, and specifically investigates a practitioner-oriented, value-chain-driven perspective for the German energy sector. In this manner, empirical theory is combined with practical experience to retrieve a comprehensive analysis upon energy-related synergies. From here, policy implications at the firm level are designed with the intention of increasing shareholder value generation.

The remainder of this chapter is structured as follows. Section 2.2 describes recent market developments, followed by section 2.3 which analyzes particular synergies of energy-related M&A. Section 2.4 considers a practical perspective upon the German energy M&A market by conducting three expert interviews with leading utility firms, which are assessed against the empirical findings in the next section 2.5. Finally, section 2.6 summarizes the main findings and indicates further areas of research.

2.2 Consolidation of the Energy Market

2.2.1 Global Energy Market Trends

The energy market consists of three different ownership types: investor-owned utilities, rural electric cooperatives, and publicly owned utilities including municipal, state, and federal providers. Investor-owned firms are the dominant market players, serving 75% of the US retail market while generating and distributing electric power. Publicly owned utilities are smaller in size but more numerous, with over 200 entities in the US. A similar split may be found across the EU (Froelich & McLagan, 2008).

Regardless of the ownership structure or energy source, the energy value chain is divided into three segments: upstream, midstream, and downstream. Each segment has specific characteristics, which may be utilized in M&A to generate synergies. The upstream sector requires scale to lower costs per customer. Thus, the focus is primarily on commodity, low margin, and high volume aspects. The midstream segment extends the focus upon scale with physical assets, to not only reduce costs, but also provide marketing and trade capabilities, which are intended to lead to asset optimization. The downstream segment is driven by the convergence between gas and electricity, in the main, as well as other energy-related services. It is moreover expected that gas will increase in relevance as an energy source; the share of electricity generated by gas was 10% in 1999, and is expected to grow to 25% in 2015. The main driver for this development is its inexpensive generation and environmental benefits (Klimchuk, 1999).

Figure 2.1: Energy value chain

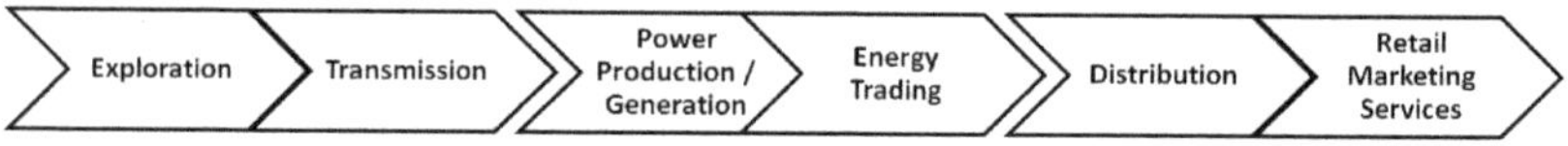

Figure 2.1 shows the energy value chain.

Owing to this development, the energy industry is in a state of transition which, in the long run, will affect the to-be-realized synergies. In particular, the regulatory framework in Europe has changed within recent years. It has been acknowledged that certain market segments, specifically energy generation and selling, have the potential to be liberalized, which implies that multiple operators can provide similar services. Other segments, i.e., transmission and distribution, still remain as natural monopolists. The traditionally parallel markets (open and constrained) were collapsed by the European Commission, and therefore, by 1 July 2007, all users had to be given the possibility to negotiate their supplier, and thus the price (Piacenza & Vannoni, 2009; Daim, Schweinfort, Kayakutlu, & Third, 2010). Moreover, this market liberalization led to increased M&A activity within the European energy market as large market players took over smaller ones. In 2004/2005, an M&A deal increase of 80% was seen; this was substantially above the global increase of 30% (Datamonitor, 2005).

In addition, the utility market convergence, which has led to an increasing relatedness among the various networks, further promotes these growth prospects (Fraquelli et al., 2004). Various energy sources are thus bundled, which changes the overall market structure by making it possible to package different energy sources such as gas and electricity. In Germany, for instance, the merger between E.ON and Ruhrgas in 2003 significantly shaped the energy environment (Granier & Podesta, 2010). A resulting question, however, is whether bundling leads to synergy effects and, if so, which effects are predominantly relevant. This, among other aspects, will be addressed in the next section.

With respect to the diversification effects of M&A, four types of energy M&A can be distinguished (Bösecke, 2009):

(1) Focused M&A between identical energy sources (i.e. electricity and electricity M&A)
(2) Convergent M&A between gas and electricity firms
(3) Concentric M&A between an electricity or gas firm with a firm dealing with another energy source (such as water)
(4) Conglomerate M&A between an electricity or gas firm with an unrelated line of business.

An investigation of US energy-related M&A reveals that firm growth manifests, yet at the expense of profitability, with the results of focused, convergent, and concentric M&A being above those of conglomerate ones (Studness, 1996). In contrast, Jandik and Makhija (2005) propose that diversified utility firms create value through investing in unrelated businesses, while focused utility firms tend to overinvest in their particular business, and thereby destroy value. This finding implies that synergy effects do not manifest. Finally, Geiger and Hoffman (1998), focusing on investor-owned utility firms, conclude an inverted u-shape relationship between diversification and performance. Thus, a moderate level of diversification is expected to benefit firm performance. Overall, the degree to which diversification drives synergy effects in this industry is thus questionable.

Furthermore, other main influential factors within the energy market include concerns about climate changes, which force governments to intervene and thus foster the reorganization of existing market structures (Flamos, Grafakos, Qikonomou, & Zevgolis, 2010). Fossil fuels, as the major source of energy consumption while providing 80% of global energy, are responsible for 60% of greenhouse gas emissions. Therefore, if the objective of a zero-carbon-emitting energy sector by 2050 in the developed world can be achieved, energy production has to change dramatically, driven by renewable energy sources – i.e. those that are free of any fuel supply constraints, produce no or only minor emissions, and are widely available (Froggatt & Levi, 2009). The EU in particular has proposed a "20/20/20" objective, with the aim of forcing a 20% greenhouse gas reduction, saving 20% of the European energy costs, and consuming 20% of energy from renewables (Stankeviciute & Criqui, 2008). Therefore, consumption patterns are expected to change, which leads to a question regarding the degree to which this change will affect the usage of energy. This is particularly important for the current research, since utility providers have to adapt accordingly, and thus will shape their respective M&A strategy around it. However, only a subset of the market is sensitive to climate changes with respect to their energy consumption. For instance, gasoline and jet fuel is highly seasonal (during peak summer travel seasons), but is not reactive to climate changes, while electricity and gas are. To measure the effects of climate-sensitive energy demand, Considine (2004) computes an elasticity measure, capturing the percentage change in consumption due to a percentage change in heating or cooling degree days. He concludes that heating degree changes have a higher elasticity for all energy sources (except natural gas), compared to cooling changes. This implies that energy consumption should shrink due to global warming, since increased fuel usage during summers is offset by reduced usage during warmer winters.

In addition, short-term industry trends reveal that utility firms tend to either focus on a wholesale or a retail strategy. Wholesalers generally concentrate on the generation of assets and gas pipelines, while retailers work on transmission and distribution functions. In the medium-term it will be essential to be well positioned on a regional, as well as national, level. In this way, not only can a critical mass be attained, but also the core strength can be fully utilized. In the long run, the ongoing transformation of the power utility market is likely to lead to three company types with distinctive added values: (1) Transaction and distribution firms, which will have a critical customer mass and strong branding, (2) generation firms which serve the commodity and wholesale market, and (3) energy services which are expanded towards non-energy related products and services, such as the telephone or internet segment (Klimchuk, 1999).

2.2.2 The Restructuring of the Germany Energy Sector

Due to the Fukushima nuclear catastrophe in March 2011, the extension of nuclear power stations agreed in September 2010 has been abolished. As a result, seven power plants had to be shut down immediately, while the willingness of large nuclear operators (i.e. E.ON, RWE, Vattenfall, and EnBW) to contribute to the so-called green fund, which is intended to fund renewable energy sources, diminished due to the fact that no extra revenues will be gained. The green fund's major focus is upon supporting the generation of wind power. The German Association of Energy and Water furthermore proposes that nuclear power shall be abolished by 2020 (BMI, 2011). Nevertheless, the current focus on wind power cannot be expected to substitute nuclear power, as can be observed by the case of Great Britain. Hence, it is far more likely that gas will increase in importance, especially with the construction of the North Stream natural gas pipeline. The pipeline, which is a joint venture between Gazprom, BASF, and E.ON, is expected to serve as a key energy source for Germany. Nonetheless, in May 2011 the first offshore wind farm in Germany was launched by EnBW, with a capacity to serve 50,000 households. Due to this recent success, more grants have been issued to EnBW by the European Investment Bank to construct another wind plant by 2013. In addition, photovoltaic solar plants are becoming increasingly popular in the German energy sector; within the first eight months of 2010, half of all photovoltaic plants installed worldwide were built in Germany. However, this rapid growth of alternative energy sources requires the expansion of power transmission grids; in particular, by 2015, 850 kilometers of new grids need to be installed (BMI, 2011).

These expected trends are supported by the Germany energy mix, exhibited in Table 2.1. As of the 1990s, oil was Germany's major source of energy, accounting for 35% of the energy usage mix (Panel B).

Table 2.1: The German energy mix

	1990	1995	2000	2005	2010
Panel A: German energy production in %					
Hard Coal	34	37	27	18	9
Brown Coal	50	40	40	40	37
Oil	3	3	3	4	3
Gas	9	14	17	14	11
Renewable	3	6	11	19	34
Remainder	1	0	2	5	6
Panel B: German energy usage in %					
Hard Coal	16	14	14	12	12
Brown Coal	22	12	11	11	11
Oil	35	40	38	36	33
Gas	15	20	20	22	22
Nuclear	11	12	13	12	11
Renewable	1	2	3	5	10
Remainder	0	0	1	2	1

Source: Data provided by the AGEB (2012)

However, the importance of oil decreased to 33% in 2010, while gas increased from 15% to 22% during the same period. Another development which fostered the relevance of gas is the decreased usage of brown coal. Furthermore, renewable energy sources increased in importance. In particular, following 2005 a substantial appreciation in renewably sources is evident due not only to the continuous decrease in nuclear energy as described above, but also governmental support programs: the usage increased by 100% between 2005 and 2010, while the production share currently represents about 34%. However, this energy source still represents only a minor source for Germany, at about 10% of the total energy usage.

Owing to the substantial amount of energy demanded and the minor generation within Germany, a substantial proportion is imported. Notably, 40% of all oil and 35% of all gas imports originate from Russia, which leads to a high dependency on this country (Frodel & Schmidt, 2009). Therefore, it is not surprising that German energy firms are shifting their focus towards Russia, as can be observed by E.ON's recent attempts to cooperate with Gazprom, Russia's leading gas utility firm. As of mid-August 2011, E.ON and Gazprom are negotiating long-term gas price contracts (Spiegel Online, 2011). Another German energy giant, RWE, has initiated cooperation with Gazprom, which intended to jointly construct power plants in Germany, Great Britain, Belgium, the Netherlands, and Luxembourg (Reuters Deutschland, 2011).

Evidently, Germany is currently facing an industry shock, which, as Harford (2005) suggests, is likely to trigger M&A waves. In particular, those shocks that require a large-scale reallocation of assets such as in Germany. Therefore, energy firms are required to adapt to the marketplace and establish a competitive position by engaging in M&A, which eventually lead to synergy gains.

2.3 Synergies in the Energy Market

2.3.1 Scale and Scope Synergies

On a theoretical level, it is suggested that regardless of the M&A type, whether it is focused, convergent, or even unrelated, scale effects manifest in the retail segment of the value chain. Through the consolidation of billing systems or call centers, costs are shared. In addition, resulting joint sales structures have the potential to lead to scale economies (Bösecke, 2009). Scope economies can also result due to shared inputs such as meter reading, accounting, and engineering services. In addition, intangible assets such as management expertise or demand forecast systems have the potential to lead to economies of scope (Sing, 1987).

Sing (1987) analyzes the realization of scale and scope synergies with respect to the gas and electrical services industry in the US in the 1980s, and examines whether combinations of these two utilities are more efficient than sole ventures with one source of energy. He concludes that the mean combination utility generates diseconomies of scope. Nevertheless, for certain combinations scope economies are present. Thus, there is potential to attain these synergies, yet the average firm is not able to do so. Furthermore, the costs for the average gas and electrical combination unit can be decreased by 7.2% if those services are provided separately. While investigating scale economies, product-specific scale economies are demonstrated for electricity, but not for gas.

A similar study conducted by Fraquelli et al. (2004) on the Italian market extends the utility focus to water utilities. In contrast to Sing (1987), economies of scope are demonstrated for firms that are smaller than the sample median. In addition, the smaller the multi-utility firm, the higher the economies of scope: cost savings range from 33% for relatively small multi-utility firms, to 12.6% for larger ones. For those firms in excess of the median size, neither economies nor diseconomies can be identified. Thus, the resulting cost savings are no longer offset by any of the factors proposed by Sing (1987), and utility firms are able to create scope economies. Concerning economies of scale, similar results are concluded. For firms smaller than the median, increasing returns to scale are identified; yet for larger firms these diminish. Therefore, small firms which engage in M&A have the potential to lower their overall costs. In addition, as already partially proposed by Sing (1987), for large firms only very minor cost advantages are identified at best; hence, recent merger waves in the EU of large multi-utility firms must be borne by aspects other than cost savings.

A detailed value chain analysis reveals additional insights. In particular, the focus is upon the generation, as well as the distribution, stage. A study by Christensen and Greene (1976) within the generation segment of the value chain in 1955 is able to prove the existence of scale economies, yet an identical investigation in 1970 is not able to do so to the same extent. It is proven that the degree of scale economies diminishes with firm size, and therefore it can be inferred that electricity-generating firms had not peaked in terms of size in 1955. During the particular time frame from 1955-1970, electricity sales in the US grew from 369 to 1,083 billion kilowatt-hour (kWh); thus, the average firm was able to triple its output. Therefore, most firms also exhausted their scale potential, and only smaller market participants are still able to achieve economies of scale (Christensen & Greene, 1976).

Within the distribution segment, *wires* and *supply* are the two main cost aspects. Wires refer to the required infrastructure, while supply refers to marketing and administrative functions, as well as the allocation of energy between the generator and customer. Overall, wire-related costs account for 59% of the total distribution costs, while supply-related costs account for the remainder. Of the total wire costs, the physical network accounts for two thirds. This particular item is a fixed cost. With respect to supply costs, the main cost items are of a variable nature, such as, for instance, customer accounts. The scale calculation reveals that economies of scale exist for the distribution as a whole, and in addition, for its component wires and supply. In particular, scale economies depend on the kWh of output, as well as on the customer density (i.e. economies of density), yet not on the territory size. However, the scale economies for wires are substantially larger than for distribution, since the fixed costs are averaged over more customers if the customer base can be increased. In addition, electricity and gas combinations in particular lower the operation costs for supply-related activities, as well as costs associated with customer accounts (Kwoka, 2005).

With respect to single-energy sources, existing evidence indicates a variety of emerging scale- and scope-related synergies along the production, trading and distribution stages of the value chain. On the production side, cost synergies can result due to common procurement activities, which eventually lead to favorable purchase conditions. Other cost synergies potentially arise during the trading stage, where human resources, as well as infrastructure, become partially obsolete due to redundancy; yet this is only the case if the M&A is within the same geographical boundary (Bösecke, 2009). Scale synergies at the distribution stage have been identified above, for instance the usage of common grid networks, as well as their maintenance or economies of density.

Piacenza and Vannoni (2009) completed an investigation of the Italian electricity market in the period 1994-2000. Overall, they report that costs increase more than proportionally, with an increase in all outputs for the median firm. Thus, for focused electricity M&A, in contrast to diversified energy-related M&A, scale synergies are identified. Moreover, scale synergies are demonstrated for vertical, as well as horizontal, aspects, and hence at a particular stage and for multiple products. On the other hand, scope economies (both vertically and horizontally) are significantly lower than the sum of the individual costs. This finding implies that firms specializing solely in serving one customer group (residual or industrial) or solely in power generation have higher costs. Therefore, this particular research identifies substantial synergy potential for electrical energy firms, and suggests that related M&A lead to synergies. Comparable scale-related findings for the electricity market are concluded by Salvanes and Tjotta (1994) for Norway; by Burns and Weyman-Jones (1996) for England and Wales; by Filippini (1996) for Switzerland; and by Yatchew (2000) for the Canadian electricity market.

On the whole, scale and scope synergies are identified for small multi-utility firms, while empirical studies cannot demonstrate these for larger companies. Moreover, scale economies are retrieved for single-utility firms, as well as for the power generation and distribution segment.

2.3.2 Financial and Operational Synergies

According to Bösecke (2009), financial, as well as operational, synergies primarily manifest during the final stage of the energy value chain – the administration stage. Here, synergies can result due to the consolidation of management boards and strategic planning departments, as well as through shared services. Particularly focused M&A have the potential to consolidate business units and thereby combine best practices and enhance operational efficiency.

Sueyoshi et al. (2009) investigate whether diversified electrical utility (gas-electricity) firms outperform electricity-concentrated ones from a financial point of view during the period between 1990 and 2004 in the US. In the first instance they show that there is a positive causality between the studied firms' financial performance and their corporate value (measured using Tobin's Q). The corporate value of multi-utility firms is higher than for single-utility firms before market deregulation; this trend reversed after deregulation. Hence, prior to deregulation, financial synergies were realized by extending the utility focus. Overall, it is thus suggested that within the deregulated European market, diversification harms financial performance, and the best strategy is therefore to focus on one utility only. This finding is in compliance with the findings on scale and scope synergies for large multi-utility firms.

Sueyoshi and Goto (2011) analyze the operational performance (productivity) of electricity firms compared to electricity/gas combinations. The investigation reveals that electricity-focused firms outperform diversified ones. Hence, there were no operational synergies present in the US for diversified utility firms. Nevertheless, the average operational efficiency of electricity-focused firms decreased from 1996 onwards, while that of diversified utility firms remained constant. This implies that deregulation policies greatly affect sole ventures in favor of multi-utility combinations (Sueyoshi & Goto, 2011). Therfore, a convergeance of the energy market can be inferred, which is further supported by the fact that bi-energy bundles between gas and electricity increasingly manifest in the market place. Granier and Podesta (2010) propose that combinations of energy sources will lead to synergies, yet these cannot by attained in the short run, following deregulation.

Furthermore, Granier and Podesta (2010) demonstrate that prior to any deregulatory activities – i.e. prior to being able to bundle products – there was no incentive to engage in any M&A activity. Independent pricing does not trigger M&A. This is in line with the studies discussed above, which were unable to identify any benefits from an energy M&A. However, once a bundle is offered to the market, all other market participants have an incentive to follow suit. Through bundling, firms attract additional customers, for instance by setting lower prices, and therefore other market participants become unable to compete. For instance, after the merger between Gaz de France and Suez, prices for bi-energy bundles were discounted to €36 compared to the individual prices on an annual basis (Granier & Podesta, 2010). However, it may be the case that such price reductions are only feasible if synergies evolve.

2.3.3 Summary of Empirical Evidence on Energy-Sector Synergies

Owing to the empirical research discussed above, synergy potential is suggested as a necessary condition for certain energy-related M&A. Table 2.2 on the following page summarizes these findings, which will serve as a guideline for the discussion of the interview-based research. The majority of empirical studies have been completed prior to the market deregulation, and thus are solely concerned with single-energy sources, especially electricity, and with scale and scope synergies. Other synergy effects, as well as bi-energy product bundles which have become popular in the recently deregulated market, have not been specifically investigated to date.

Table 2.2: Synergies in the energy sector: an overview

Study	*Utility Focus*	*Synergy Objective*	*Timeframe*	*N*	*Findings*	*Comment*
Christensen and Greene (1976) Focus: US	Power generation	Scale Synergies	1955-1970	124	Scale economies diminish with increased firm size	/
Sing (1987) Focus: US	Gas/electricity utilities	Scale/Scope Synergies	1981	108	Mean utility firm has diseconomies of scale/scope	Certain combinations display synergies
Salvanes and Tjotta (1994) Focus: Norway	Electricity utilities	Scale Synergies	1988	91	Scale synergies only for small utilities	/
Burns and Weyman-Jones (1996) Focus: England/ Wales	Electricity utilities	Scale Synergies	1980-1993	12	Evidence for scale economies for all utilities	/
Filippini (1996) Focus: Switzerland	Electricity utilities	Scale Synergies	1988-1991	39	Economies of density for small/ medium utilities	No scale synergies for large utilities
Yatchew (2000) Focus: Canada	Electricity/ water utilities	Scale/Scope Synergies	1993-1995	81	Scope synergies are suggested for small utility firms	/
Fraquelli et al. (2004) Focus: Italy	Gas/electricity/ water utilities	Scale/Scope Synergies	1994-1996	90	Small utility firms display economies of scale/scope	No synergies for large multi-utilities
Kwoka (2005) Focus: US	Power distribution	Scale Synergies	1989	436	Scale economies are conducted	/
Piacenza and Vannoni (2009) Focus: Italy	Electricity	Scale Synergies	1994-2000	25	Vertical and horizontal scale synergies	/
Sueyoshi et al. (2009) Focus: US	Gas/electricity utilities	Financial Synergies	1990-2004	104	Synergies vanished for multi-utilities after deregulation	/
Sueyoshi and Goto (2011) Focus: US	Gas/electricity utilities	Operational Synergies	1990-2004	104	No synergies for multi-utilities	/

2.4 Methodology and Findings

2.4.1 Interview Method and Rationale

If the transaction premium is utilized as the only tool to infer the existence of synergies, practical evidence proposes that these exist for all energy-related M&A. On average, the premium for purely electricity-focused M&A in 1999 was 22%, while for a mixed-utility merger between electricity and gas it was

27% (Klimchuk, 1999). However, the empirical research discussed above shows only minor synergies for specific M&A transactions; thus, the additional perspective of practitioners will shed light upon this discrepancy. Therefore, the present research intentionally conducts a clinical study in order to identify unique aspects which do not *average out* though a larger sample size.

Additionally, in the course of daily business, managers are confronted with obstacles or incentives that empirical research has not explicitly focused on. These obstacles may include, for instance, Roll's Hubris Hypothesis (Roll, 1986), agency conflicts (Mueller, 1969), diverging incentives between the management and firm owners (Painuly, 2009), or opportunity costs (Edlin & Stiglitz, 1995). However, such issues may significantly shape the perception of synergies; thus, disregarding them potentially results in misleading conclusions. Therefore, interviews as a supplementary research instrument are most suitable to extend and aid to the understanding of the existing energy sector synergy research. Thus, the intention is to incorporate, as well as to expand, the field of synergies according to the managerial and, to a certain degree, behavioral perspective of the decision makers involved. In line with this goal, the interviews used open-ended questions to enable the experts not only to direct their focus towards their perceived importance of synergies, but also elaborate in order to obtain rich detail. Furthermore, prior empirical studies have unanimously focused on single-synergy effects, while the interview research methodology makes it possible to additionally investigate the interdependency and relative importance of various synergy sources. Moreover, the extensive practical experience of the interviewees increases the expressiveness of their opinions and facilitates inferences, in combination with existing empirical findings, regarding actual perspectives upon the generation of synergies and correspondent shareholder value.

In addition to using open-ended questions, the interviews followed the outline below (see Figure 2.2), with the objective of identifying drivers of shareholder value generated by synergies. Therefore, the interviews were split into three sections with respect to each synergy source. The focus of each section was to identify:

I. The perception and relevance of the synergy source within the energy sector, and
II. The primary value-driving characteristics.

With this in mind, the literature review above serves as the basis for the discussion.

Figure 2.2: Interview structure

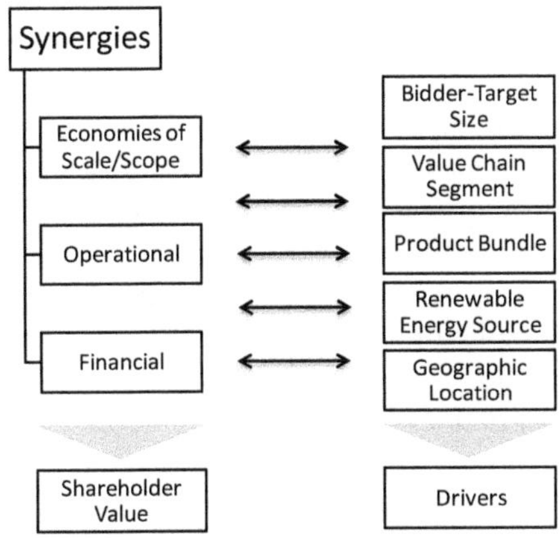

Figure 2.2 is an own illustration that outlines the structure of the open-ended expert interviews.

The interviews were conducted on 30 August, and 9 and 13 September, 2011, with three managers of the three leading, publicly traded energy providers in Germany. The first interviewee, who is currently serving as the head of controlling and M&A valuation, has more than eight years of relevant work experience. Prior to his current position, he was employed as an M&A project manager for five years, and served two years as the assistant to a chief financial officer. The second interviewee, who has five years of M&A experience, is employed as an M&A project manager. Prior to this position, the expert served as an M&A speaker for three years. The third industry expert has worked for leading consulting firms with a focus on the financial sector and M&A advisory for more than five years, after joining a German utility firm in 2008. Currently, the interviewee is employed as a group speaker for group planning and controlling. The interviewees thus have more than 17 years of combined experience, and followed as well as valued multiple M&A transactions in the field of energy. For disclosure reasons, the interviews are discussed anonymously, without indicating any firm-specific information.

2.4.2 Findings

Overall, the experts' statements concerning the multiple research areas had a high degree of uniformity. Thus, it was not deemed necessary to obtain additional, relevant perspectives by conduction further interviews. On the whole, the generation of synergies is not described as the primary goal of the M&A activity; rather the strategic reorientation is placed as the initial objective. Frequently, the M&A strategy is to increase the market presence and thereby strengthen the market power. Synergies are described as a *nice-to-have* factor, while their role is twofold. They are, in fact, essential for the bidding price calculation, and thus for premium calculations. On the other hand, they are a critical aspect that can be released to internal bodies, as well as investors, to support the M&A.

Financial synergies appear to be the simplest to attain, and are not specifically generated due to any energy-related activity, but owing to the financial transaction. One interviewee stated that he was not aware of any transaction in which financial synergies, besides transaction-related ones, had evolved at all. Therefore, it may be implied that these particular synergies demonstrate a side effect, and are not specifically taken into consideration in M&A.

Operational synergies are directly associated with cost savings. These can especially result if an already established business unit is increased in size, and they particularly manifest in the distribution and retail segment of the value chain. Thus, the driver of this synergy is the number of customers. One interviewee argued that the cost savings especially manifested in the smaller firm of the merging entity, while another interviewee perceived that the operational synergies can be maximized if the merging entities are of equal sizes, and thus have an equal rights allocation. Under these conditions, organizational structures, as well as the number of customers, correspond, which enables the most efficient integration and synergy realization. However, operational synergies are also identified as being the most difficult to realize within the German market. According to one interviewee, roughly one third of the intended and communicated magnitude is actually realized. An explanation for this can be redirected to the organizational structure of most German energy firms. The significant market players, which were formerly state-run, are still subject to highly bureaucratic structures. Therefore, the implementation and reorganization of organizational procedures is difficult and time-consuming. For instance, reducing the headcount is supposed to be a common operational synergy, yet German firms in particular are struggling to do this due to the influence of powerful employee work councils.

Economies of scale and scope are indicated to be especially present in certain areas of the value chain, i.e. within energy trading and purchase aspects. Purchasing power can be built through increased size, and thus an increased amount of kWh, which enables favorable purchase conditions and reduces costs. Moreover, if M&A are able to increase the market power, margins can be increased, which would also serve as a synergy effect. Hence, the latter aspects of the value chain can also be targeted to generate synergies. In addition, scope and scale economies manifest within interdisciplinary functions. However, these synergies are described as being less substantial compared to those in other industries, since their potential is not as large. The existing networks can only be leveraged to a certain degree, and thereby cap the synergy potential. Therefore, if the territory size is increased through the merger, new network grids have to be acquired or built, and hence, newly gained customers cannot be fully utilized to spread the fixed costs.

With respect to product bundles, the gas/electricity bundle is expected to demonstrate the highest synergy potential, since these energy sources illustrate the same features, such as consumption-dependency and price-sensitive customers. Thus, operational synergies can be attained. In addition, customers within this segment – so-called dual-fuel customers – are those that demonstrate the highest profit potential, since two products are sold in one instance. Thereby, two margins are gained while associated costs occur only once. This, however, leads to intense competition in the gas/electricity market segment. Furthermore, this market is expected to grow; thus, the convergence of electricity and gas can reasonably be projected to increase, which should lead to increased synergy effects across markets.

Bundles with renewable energies serve as a pulling factor, rather than as a synergy-generating element. Thus, customers will be attracted by additional bundle offers with renewable energies, yet potential synergies cannot be redirected to the regenerative nature of the energy. Nevertheless, German energy firms are actively pursuing M&A with regenerative energy sources in order to acquire the most recent technology for their energy generation portfolio, and to be able to be part of the knowledge generation process. Since the customer's willingness to pay in excess due to the renewable energy generation is low, no scale can be attained, which potentially enables synergies. As has been elaborated above, the synergy source is the increase in customers, and this has not manifested in the market as of yet.

Another perspective indicates that no bundle reveals any superior synergy potential, but the firm's focus on, for instance, exploration or distribution, as well as the territory size, accounts for synergies. Therefore, in contrast to the views found by prior industry experts, traditional and renewable multi-utility combinations have an identical synergy potential.

In addition, all interviewees describe the geographical distance, and hence the consequent cultural distance, as important for M&A success and synergy generation. A precondition for synergies is the integration of departments, and industry experts propose that a closer cultural distance leads, on average, to higher synergies. However, the *distance* at which the synergy generation is affected is perceived differently. One interviewee argues that M&A outside the boundaries of Germany is likely to affect synergies, while another expert perceives that only M&A outside the EU have the potential to do so.

2.4.3 Discussion

Empirical research documents solely minor synergies, yet premiums (reflecting synergies) in the order of 22% for mixed-utility mergers (Klimchuk, 1999) are paid; hence, managers perceive synergies differently; a final comparison assessing the empirical research perspective against the managerial perspective with respect to synergy sources and value drivers is presented in Tables 2.3 and 2.4.

Table 2.3: Importance of shareholder value effects from synergies

Study	*Scale and Scope Synergies*	*Operational Synergies*	*Financial Synergies*
Christensen and Greene (1976)	Medium importance	-	-
Sing (1987)	Low importance	-	-
Salvanes and Tjotta (1994)	Medium importance	-	-
Burns and Weyman-Jones (1996)	High importance	-	-
Filippini (1996)	Medium importance	-	-
Yatchew (2000)	High importance	-	-
Fraquelli et al. (2004)	High importance	-	-
(Kwoka, 2005)	High importance	-	-
(Piacenza & Vannoni, 2009)	High importance	-	-
Sueyoshi and Goto (2011)	-	Low importance	-
Sueyoshi et al. (2009)	-	-	Low importance
Interviews	High importance	Medium importance	Low importance

The interviewees conclude that scale and scope economies are the primary synergies arising regardless of the M&A, be it focused or unrelated. This finding is in support of empirical research which demonstrates scale and scope economies for certain energy-related M&A. The synergy sources, i.e. those merger characteristics driving the value effect, are described, according to the industry experts, in a similar way to those identified within the US market (Kwoka, 2005), namely in terms of increased customer density, and not increased territory size. A distinction between these features is of significance, as territory size may increase while customer density does not; hence, M&A officials need to examine the population within potential acquiring regions. Moreover, empirical research evidence and managerial experience confirms the presence of scale and scope synergies, particularly within the power generation segment of the value chain. In addition, financial synergies are of minor relevance, and thus the empirical research, which was not able to demonstrate any financial synergies (see Sueyoshi et al., 2009) in the US market, is supported within Germany. No financial synergies may be traced back to any energy sector chain link.

Table 2.4: Synergy effect drivers

Study	***Value Effects***	***Interview***
Christensen and Greene (1976); Salvanes and Tjotta (1994); Fraquelli et al. (2004)	Synergies diminish w/ firm size	Synergies can be maximized with similar firm size
Sing (1987)	Increased quality diminishes synergies	-
Yatchew (2000)	Product bundles drive Synergies	Product bundles drive synergies
Burns and Weyman-Jones (1996); Kwoka (2005)	No. of customers and kWh drive synergies	No. of customers and kWh drive Synergies
Piacenza and Vannoni (2009)	Related M&A drive synergies	Unrelated product bundles experience increased synergies
Sueyoshi et al. (2009)	Deregulation decreases synergy potential	Cross-sectional mergers enabled through deregulation increase Synergies
Sueyoshi and Goto (2011)	Deregulation increases synergy potential	Cross-sectional mergers enabled through deregulation increase Synergies

The perspective on operational synergies is twofold. Energy-related M&A have the potential to generate them, yet their *ex ante* measurement and *ex post* generation is questionable. A significant proportion of this synergy type is characterized as best practice approaches and subsequent value (i.e. synergies) results from their transfer to the bidder or target firm. Most frequently, according to the interviewees, the bidder firm, being the more powerful party, transfers its knowledge and implements its operational procedures. However, these aspects, especially their monetary value effect, can hardly be measured *ex ante*, and thus operational synergies are somewhat ambiguous. This finding therefore serves as an attempt to explain the results obtained by Sueyoshi and Goto (2011), who were not able to prove operational synergies for multi-utility firms. Nevertheless, the interviewees do not perceive operational synergies as irrelevant as these can emerge given the respective environment; this is in contrast to the empirical studies, which are based on stock market data. Moreover, the ambiguity of operational synergies hints towards the political aspect of synergies, which is an essential facet that has not been considered by prior studies based on capital markets. The interviewees mention that synergies are a means by which to support M&A decisions, and therefore higher synergies are expected to increase the probability of the realization. Thus, if certain approval processes are necessary, synergies might be computed towards a higher magnitude, especially with respect to ambiguous aspects such as operational synergies. This conclusion may also demonstrate the presence of the Hubris Hypothesis (Roll, 1986) within the German energy market. Managers in the energy sector seem to be overconfident about the realization of operational synergies; hence, communicated synergies are possibly not realized. As these are, however, most likely priced in the transaction value, shareholders will, under such circumstances, experience a loss from their non-realization.

This political aspect also reveals its relevance by considering the efforts towards renewable energy sources. The experts indicate that only minor synergies are possible due to the renewable nature of the energy, yet M&A are pursued between black and green utility firms. Communicating synergies in the light of this process may thus serve as an efficient tool to support the M&A decision, yet their degree of implementation is again questionable, which prevents empirical research from demonstrating them and again highlights the necessity of the interview-driven research design. Thus, it appears that synergies also serve as a tool to fulfill personal goals under the deregulated environment. Moreover, M&A in the German energy market, as the experts describe, apparently has the potential to evolve market power; this contradicts the findings of research by Fee and Thomas (2004).

With respect to product bundles, the expected convergence of gas and electricity predicted by empirical studies can be partially confirmed. Thus, the favorable development of the deregulation for utility bundles, as expected by Granier and Podesta (2010) for the US market, also manifests in the EU. One of the experts in the present study further indicates a higher synergy potential for gas/electricity bundles in contrast to other bundles, which is yet not in accordance with empirical findings by Fraquelli et al. (2004). The reasoning for this discrepancy may be that due to intense price competition in the German and also the European market, emerging synergies are frequently used to lower consumer prices, and thereby the previously gained benefits diminish. Thus, market pressure may force firms to fully transfer the value generation (i.e. synergies) towards the consumer. Other experts support the view that all bundles create a similar synergy potential, while also considering renewable energy sources. Therefore, whether certain utility combinations demonstrate higher synergies cannot be fully concluded.

Lastly, the interviewees indicate a discrepancy between planned, and thus communicated, synergies and their actual implementation, which decreases the realized shareholder value; the transaction price incorporates communicated synergies and the failure to realize these decreases value. Nonetheless, the potential to generate synergies is given throughout the value chain, as indicated.

2.4.4 Implications

Overall, the following policy implications can be drawn at the firm level. M&A are designed to maximize shareholder wealth generation and particularly evolve with regards to the deregulated European energy market. Synergies in general do not increase with increased territory size; hence, value-generating M&A should seek targets that maximize the number of customers and also widen their focus towards other energy sources. Besides these resulting synergies, a diversified utility portfolio also serves a risk-reducing role, which consequently also increases the perceived shareholder value. Secondly, empirical, as well as managerial, research assigns the highest synergy potential to scale and scope economies. However, these predominantly only evolve if the merger leads to the creation of product bundles (i.e. an unrelated M&A), and moreover increase with a higher geographic proximity. Hence, an ideal M&A should:

(1) Add a novel energy source to the product portfolio, and
(2) Include a target firm located within the same region.

The industry experts unanimously perceive mixed bundles to share operational costs at the maximum level, while increasing revenues. The latter is the driving value-generating force, as related mergers increase revenues to a lesser extent. Therefore, contradicting empirical capital market-based studies, operational synergies are relevant. Thirdly, the payment of an increased transaction premium solely owing to expected operational synergies is nevertheless likely to lead to a loss of shareholder value. The realization of this particular synergy source is highly ambiguous, and thus can partially be predicted *ex ante*. Consequently, decision makers should not allocate too much weight to their importance in terms of merger benefits. They should at least include a kind of precaution discount for expected operational synergies in order to prevent loss for their shareholders. Furthermore, mergers with renewable energy sources do not have an increased synergy potential. Renewables are highly supported by government institutions, and are perceived as *the* future growth market, which fosters utility firms to pursue M&A within this market sector. Nonetheless, firms should use caution when basing the transaction premiums upon emerging synergies generated through these transactions, and should rather base potential premiums on future market growth and an eventually increased market share. Otherwise, a loss in shareholder wealth appears to be likely.

2.5 Conclusions

This chapter intends to advance the empirical discussion on energy-related synergy effects, particularly in the German market. Initially, to establish a theoretical understanding, synergies within the energy sector are discussed. In a subsequent study, this discussion is supplemented via interviews with industry expert. The analysis reveals that the industry's unique structure highly influences the synergy potential. The European energy market is in a state of transition towards liberalization. All energy providers are thereby enabled to acquire or merge across markets, which also enables single firms to provide multiple services, such as bi-energy product bundles. In addition, the government-demanded transition towards renewable energy sources requires energy providers to widen their product portfolio, and a common means by which to do this is through M&A.

The investigation on scale and scope economies demonstrates that while considering the whole energy value chain, only minor synergies exist, if they exist at all. Scale and scope economies are only demonstrated for small utility firms. However, a more detailed examination shows that certain segments of the value chain have synergy potential; namely, distribution and generation. These empirical findings are supported by the industry expert interviews. Operational synergies, the second most relevant synergies according to the interviewees, are mostly present in the distribution and retail segments of the value chain. Furthermore, it is indicated that the size relation between bidder and target matters; however, perspectives vary significantly, and no final answer concerning the target or bidder size can be concluded.

Finally, financial synergies are perceived to be less relevant in the energy sector, in comparison with other industries. Their generation, according to the expert interviews, is not unique to any energy-related M&A, and empirical research concludes that diversification towards other utilities actually harms the financial performance (Sueyoshi et al., 2009).

To further increase the expressivness of synergies during energy-related M&A in the EU, the quantitative analysis can be advanced based on the implications of the expert interviews. In this light, it is relevant to consider how the capital market evaluates the communicated synergies, since their measurement is highly ambigous; a gap between communicated and implemented synergies has been proposed, in particular due to the political behavior of managers. Hence, the question evolves as to whether the capital markets discount the communicated synergies. In addition, the varying effects of focused and unrelated M&A should be analyzed, while the factors contributing to the magnitude of energy-related synergies should be explored in more detail. These include the cutural distance or size relation between the bidder and target firm.

3 Synergy Communication Effects in European Energy Markets[2]

3.1 Introduction

Berkovitch and Narayanan (1993) demonstrate, via a cross-industrial investigation of M&A motives, that synergies are the primary objective of M&A; however, the voluntary disclosure of synergy forecasts has so far received little attention, especially as there is no binding obligation to communicate expected synergies in an M&A transaction. Voluntary disclosure studies have thus far primarily focused on accounting-related releases (Lev & Penman, 1990; Baginski et al., 1994; Lang & Lundholm, 1996; Baginski & Hassell, 1997; Healy et al., 1999; Bowen, Davis, & Matsumoto, 2002; Hutton, Miller, & Skinner, 2003). This chapter extends this field of research by two essential aspects. Initially, an analysis is conducted regarding the synergy communication effects at the merger announcement on target and bidder share prices within the European energy sector. This complements the evidence on voluntary disclosure generally, and synergies at the merger announcement specifically. Secondly, a long-term investigation is completed to infer the long-term valuation effects of synergy communication transactions. This investigation is of particular importance for the European energy sector, as not only has its M&A landscape changed, as the prior chapter shows, but also its rate has increased significantly, following market deregulation. As this general trend is expected to continue, the following study thus shows the capital market evaluation of these deals, and investigates their respective value creation potentials. In particular, the value potential of additional, voluntary information disclosure is investigated.

Another objective of the study, however, is to move beyond pure measuring and assigning profits to transactions by explaining more fundamental value creation factors. The drivers of increased announcement returns are particularly important for an M&A's value generation, since these returns not only reflect the investor's perception of the deal, but also affect the completion probability (Luo, 2005). Therefore, if the communication of synergies is an efficient tool to increase announcement returns, it can be implied that they also serve as an instrument to increase the likelihood of completion. The underlying rationale is

---

2 I am grateful to Prof. Dr. Dirk Schiereck for helpful insights and comments. An abbreviated version of this chapter has been published in the International Journal of Energy Sector Management, Volume 6, Issue 4, pp. 534-557.

that M&A in general are subject to a high degree of asymmetric information (Moeller et al., 2007; Officer et al., 2009). Reducing these asymmetries through voluntary information disclosures, such as synergy forecasts, results in favorable investor reactions. However, the question of whether these positive returns remain constant beyond the announcement windows remains open. This chapter, considering a dual foci (short-term vs. long-term returns) is able to respond to this issue.

Thus far, synergy research has mainly considered a cross-industrial perspective to infer general implications. Dutordoir, Roosenboom, and Vasconcelos (2010) demonstrate that acquiring firms in the US that disclose synergies are attributed with favorable stock price reactions, and hence firms use synergy disclosure as a value-signaling tool. Bernile and Bauguess (2010) also report positive stock price effects for the US owing to synergy disclosure decisions. However, since significant differences are found for varying industries, the need for a detailed industrial analysis due to the diverging industry characteristics of the energy sector arises; the present study responds to this need. In particular, due to these idiosyncratic industry features, cross-industry investigations have the potential to result in misleading industry sector conclusions.

To address the sector-specific relevance of synergy forecasts, the energy sector is highly suitable as this industry is characterized by a high ratio of voluntary disclosure and similar business models between the companies in the sector. If it is not possible to detect synergy disclosure effects in such a sector, it is hard to imagine to find them at all. To further increase sample homogeneity with respect to general market conditions, this research takes a geographic focus on the European energy sector. This sector is especially relevant due to its high M&A rates owing to government-enforced liberalization efforts. Thus far, no study exists that conducts a synergy disclosure effect for the European market. Hence, the first study within this research may demonstrate similar disclosure effects to those retrieved for the US market, but for a novel marketplace, and, more importantly, within a distinctively different regulatory environment. Moreover, the energy industry has a significant synergy potential. Bernile and Bauguess (2010) show that the synergy disclosure rate for the US energy sector is 30.7%, while for all industries combined it is only 22.7%. Hence, synergies are particularly relevant for this industry, and thus can be expected to exhibit a more dominant role, as compared to a general industry observation.

The remainder of the chapter is structured as follows. Sections 3.2 and 3.3 conduct a review of the European energy market, as well as voluntary pre-merger information disclosure effects in general, and merger announcement returns. Section 3.4 identifies the research questions, while section 3.5 presents the research model. Section 3.6 shows the announcement effects of synergy disclosure, and Section 3.7 develops recommendations for energy managers seeking M&A. Section 3.8 concludes the chapter.

3.2 M&A Opportunities in the European Energy Market

The European energy market has undergone extensive deregulation in the past decade; on the one hand this has led to various value-creating opportunities, yet on the other it has also created challenges for European energy firms. The European Commission initiated this process with Directive 96/92/EC in 1996, while it is acknowledged that certain market segments, especially power generation and selling, can be liberalized, which thus enables energy firms to merge across the value chain, as well as into other energy sectors. It therefore seeks to break up monopolies while unbundling existing networks. Other chain links, i.e. transmission and distribution, remain natural monopolies and continue to be regulated. In particular, the third energy package issued by the European Commission in 2007 fostered and expedited the transition (EC, 2007). One of the key opportunities arising from this is the possibility to create product bundles, i.e. serve existing customers with multiple products, such as gas and electricity. The energy markets have already started to converge towards these bundles, as Fraquelli et al. (2004) show; however, the question of whether energy firms can thereby create synergies (i.e. value) remains open. This is particularly relevant for the European market, as research on the US energy sector by Studness (1996) shows that firms grow following diversification (i.e. a bundling strategy), yet this growth manifests at the expense of profitability. In contrast, Jandik and Makhija (2005) find, again for the US market, that energy firms investing in related lines of business also destroy value, as they tend to overinvest. This further implies that no synergies evolve during focused M&A. The following research responds to these issues and investigates the value potential of M&A in the European energy market and respective capital market valuation, which is of significant importance as gas is expected to increase in relevance to up to a 25% share of total electricity generation, by 2015 (Klimchuk, 1999). Therefore, the M&A rate can be expected to remain high.

Other highly influential factors are climate concerns and the respective growth of renewable energy sources, as eleborated in chapters 2.2.1 and 2.2.2. Regulatory authorities in the EU are pushing towards this growth with the 20/20/20 objective explained above (Stankeviciute & Criqui, 2008). This subsequently forces the reorganization of existing market structures, while consumption patterns are thus expected to change as well. This is especially relevant for the present research, as European energy firms have to adopt their strategy accordingly and foster growth towards these energy sources. A common means by which to grow is through M&A, and therefore the respective M&A strategy has to be adopted.

In addition, it may be observed that energy firms consider different geographic M&A strategies following the market liberalization as a subset of the market attempts to develop towards a pan-European strategy (i.e. grow through cross-border M&A), while others intend to become national champions (Verde, 2008). While there are rationales for each strategy, true value is solely generated if synergies evolve, and, more importantly, if the capital markets are able to perceive these. Thus, energy firms are essentially confronted with two issues:

I. The opportunity to create synergies through mergers, and

II. The benefits of disclosing these.

3.3 M&A Announcement Effects: Literature Review

3.3.1 Voluntary Pre-Merger Information Disclosure

Thus far, evidence on voluntary disclosure has focused on accounting-related releases. Baginski and Hassell (1997), for instance, examine the precision of earnings forecasts, while concluding that the information value of these particular forecasts increases with the number of analysts following the firm. Moreover, it is demonstrated that earnings forecasts in general have the potential to influence the share price of publicly traded firms (Baginski et al., 1994). Other voluntary disclosure research is related to extended press releases or conference calls clarifying and elaborating earnings announcements (Bowen et al., 2002; Francis, Schipper, & Vincent, 2002; Bushee, Matsumoto, & Miller, 2004; Kimbrought, 2005). All studies indicate that there is an influence on the respective response group, and thereby demonstrate the effectiveness of voluntary disclosure.

In terms of M&A, Lipin and Sirowner (2003) argue that the communication strategy can be the decisive factor for the success or failure of the deal. Consequently, it is especially important to provide quantitative forecasts, not only to enable investors to track post-merger performance, but also to evaluate the deal. By failing to indicate vital information, uncertainties are created. If the market is aware that private information exists that is not communicated, negative content is implied. Thus, the disclosure decision can have a substantial impact on the post-announcement performance of the share price (Dye, 1985). The importance of quantitative forecasting is further highlighted by Hutton et al. (2003), who show that while stock price reactions to earnings forecasts without supplementary quantitative information are close to zero, those with quantitative forecasts result in positive returns of on average 3.84%.

Nevertheless, disclosure behavior in takeovers differs from routine situations, and more recent research is thus concerned with respective share price effects. Erickson and Wang (1999), along with Louis (2004), investigate the effects of an upward manipulation of earnings announcements prior to a mergers, and demonstrate a positive correlation. Jo and Kim (2007) analyze the effects of the disclosure frequency in seasonal equity offerings and report a positive correlation with post-issue performance which hints at a reduction of information asymmetries prior to the transaction. This finding is confirmed by Lang and Lundholm (2000); however, the increase in disclosure frequency has obviously created a hype, and the respective stock price declined disproportionally in the post-announcement phase.

The following research differs from earlier studies since the disclosures under consideration are specifically directed at an event (i.e. the M&A), and are not of a general nature, such as an earnings forecast around the M&A announcement. Kimbrought and Louis (2011) take a comparable focus by investigating the effects of conference calls at merger announcements. They demonstrate that those bidders which hold calls at the merger announcement disclose a greater degree of information and emphasize forward-looking details. Managers in general, especially in stock-based transactions, have an incentive to disclose additional deal-related information, since the completion of the merger is highly dependent on the post-announcement stock performance of the bidding firm. Furthermore, the incentive for conference calls increases with the size and complexity of the transaction, where the merger motivation, as well as the benefits, are less clear to investors. Overall, Kimbrought and Louis (2011) demonstrate that merging firms that utilize conference calls experience higher announcement returns compared to those who fail to do so. Thus, the additional information content released to investors appears to increase the announcement returns.

3.3.2 Synergies as Merger Motive, and Disclosure Effects

Three main takeover drivers motivate managers for M&A transactions: hubris, agency effects, and synergies (Berkovitch & Narayanan, 1993). The Hubris Hypothesis initiated by Roll (1986) suggests that managers are unable to correctly value target firms and are thereby likely to overpay and destroy shareholder wealth. Hence, these takeovers are driven by the management's overconfidence in their ability to extract gains. The agency effect proposes, again, that through the merger it is not the shareholder's wealth that is increased, but the management's wellfare. It is implied that the compensation is a function of the firm size, and thus managers are likely to lower their hurdle rate to take on additional investments which increase the firm size at the expense of their shareholders (Mueller, 1969). Finally, synergies result if the merging firms' resources are combined in an efficient manner that leads to economic gains. Hess (2010) shows that M&A has the potential to increase the combined entities' efficiency, and hence lead to synergies. Moreover, a takeover can pursue multiple objectives; thus, these three motives are not mutually exclusive. Yet, Berkovitch and Narayanan (1993) show that the realization of synergies is the primary M&A driver. Zhang (1998) confirms this general proposition for the US banking industry, and further highlights the importance of synergies during takeovers.

From the shareholder's point of view, synergies during M&A are an indicator of increased profitability of the newly formed entity, which eventually increases the firm's value. Moreover, an increase in the firm value translates into the stock price of the merging firms, which should appreciate. However, information on the merger motivation itself, i.e. whether synergies exist, must not be disclosed by bidders. The necessary ad-hoc information only needs to incorporate information that has the potential to influence the stock price. In the case of M&A, it is hence sufficient to indicate the transaction without any further details. The most frequently used vehicles to do this are press releases. Nonetheless, firms can voluntarily disclose additional information such as the motivation of the merger and, if applicable, a quantitative forecast such as a synergy estimate. It can be expected that additional pre-merger information indicating a value increase should be appreciated by the market.

Bernile and Bauguess (2010) confirm this notion by collecting synergy estimates of US mergers between 1990 and 2005. They demonstrate that the market's reaction to those mergers was favorable; in particular, the capital market's evaluation of M&A with communicated synergies was higher, and increased in value with higher synergies. However, the study also infers that the variation across industries is high, which requires unique industry-related studies. This proposition is underlined by Jaggi (1978). Thus, to increase accuracy, as well as expressiveness of synergy disclosure effects, the following research focuses solely on one industry: the energy industry.

3.4 Research Question

This sub-chapter firstly investigates the effects of synergy disclosures regarding the M&A transactions on shareholder wealth, while considering a short-term, as well as long-term perspective. Secondly, the characteristics of synergies within the European energy market are identified; their relative size compared to other industries, as well as their value drivers, are discussed. Overall, five hypotheses are proposed.

***Hypothesis 1*:** Short-term value effects for bidding energy firms increase if a synergy forecast is disclosed at the M&A announcement.

Many studies on wealth gains from M&A conclude that the bidding firm's stock price frequently decreases on merger announcements (Morck, Shleifer, & Vishny, 1990; Moeller, Schlingemann, & Stulz, 2005). One of the major driving forces thereupon is asymmetric information concerning the merger rationale (Moeller et al., 2007; Officer et al., 2009). The present study therefore expects that if bidding firms decide to communicate a synergy estimate at a merger announcement, information asymmetries will be reduced. Hence, if firms communicate expected synergies, investors are informed about the merger motive, as well as the implied wealth gains, which can serve as a value-creation signal for rational market participants. Increased confidence about the profitability of the transaction can hence be implied, which should translate into the stock market's reaction. It has already been demonstrated that other elements that reduce information asymmetry, such as conference calls (Kimbrought & Louis, 2011), affect the announcement returns positively, and a similar effect for synergy estimates is expected to be found here.

Hypothesis 2: Short-term value effects for target energy firms increase if a synergy forecast is disclosed at the M&A announcement.

Similar to hypothesis 1, it is also expected that abnormal returns for target firms will increase if the bidder's management discloses a synergy estimate. Following the arguments presented here, a reduction of information asymmetries is likely to result in favorable stock price reactions. However, in contrast to the bidding firm's announcement returns, targets are frequently subject to significant positive reactions (Andrade, Mitchell, & Stafford, 2001; Brunner, 2004). Therefore, both sets of targets (i.e. targets with and without synergy disclosure) are anticipated to have positive announcement returns, yet those transactions with a synergy disclosure are expected to see an even more marked increase in these returns.

***Hypothesis 3*:** Relatively higher communicated synergies lead to higher short-term abnormal returns for bidding and target firms which disclose synergies at the M&A announcement.

Even though managers can most accurately value synergies due to their insider perspective and their firm, as well as industry, knowledge, investors may not take their word for it (Houston, James, & Ryngaert, 2001). The question thus arises regarding the degree to which management estimates are credible, especially as it has been shown in the fist chapter of this dissertation that synergies also serve as an instrument by which to pursue political objectives. As long as managers have a long-term horizon and intent to creat value by acquisitions rational investor should acknowledge these efforts. Thus, it is expected that the investors' reaction to the M&A will be reflected by the amount of communicated synergies. The degree to which these parameters correlate will be investigated, and this will lead to inferences concerning potential discounts, and thus management credibility. To ensure the comparability of the communicated synergies across the sample, the synergies are discounted to their present value and scaled according to the market capitalization of the bidder and target firm. Moreover, this particular analysis will also confirm the notion that the disclosed synergies are perceived as incremental information.

Hypothesis 4: Long-term value effects for bidding energy firms are positive,following a synergy disclosure.

Most prior long-term post-merger studies conclude a loss in shareholder value for bidding firms. These negative returns are retrieved for regulated and unregulated industries (Ferris & Park, 2001). However, none of these studies conduct their long-term analysis using solely synergy-disclosing firms, which are expected to decrease information asymmetries, and thereby not only increase their value perception, but also create long-term value. Therefore, if the short-term analysis concludes increased positive announcement returns for synergy-disclosing bidding firms, yet these returns decrease in the following years, it may be implied that the expected synergies were not sufficiently realized. Hence, the present study also tests whether potential short-term positive reactions are due to investor irrationality or overly optimistic management forecasts. If this is found to be the case, a negative performance should be observed in the long term. Nevertheless, as positive short-term effects are expected, they should also be consistent in the long-term, particularly since emerging synergies are not a one-time event, but manifest each year and hence generate a long-term value.

***Hypothesis 5*:** The relative magnitude of the disclosed synergies correlates with the relatedness of the transaction.

Healy, Palepu, and Ruback (1997) predict synergies to be highest in related acquisitions, which also motivates managers to use the degree of similarity as an instrument to forecast synergies (Lubatkin & Srinivasan, 1997). Therefore, when the merging entities are more alike, it is expected that the magnitude of the realizable synergies increases. However, there are several different sources of similarity, i.e. culture, business, or size, while the effects of each, especially in the European energy sector, are hard to predict. All of them are expected to be positively correlated to the magnitude of disclosed synergies.

Cultural relatedness is measured using a dummy variable that assigns all national M&A with 0 and cross-border ones with 1. Business relatedness is identified through the firms' SIC codes, while firms that share an identical code are considered to be related. Again, dummy variables are assigned; transactions with business relatedness are attributed with 0, and the remainder with 1. Size relatedness is measured according to the merging firms' sales at the year of the announcement. To ensure the comparability of the data set, a ratio of target sales to bidder sales is computed.

3.5 Data and Methodology

3.5.1 Data

The sample of this chapter consists of all completed energy-related M&A that occurred between 1 January, 1998 and 31 December, 2010 in Europe; to identify the respective M&A, SIC codes 13 and 49 are utilized. To construct the sample, the Thomason Financial SDC database is utilized while applying the following selection criterion:

- Minority deals are excluded
- A minimum transaction value of 1 million USD has to be present
- Only purely public transactions are considered, i.e. target and bidder firms have to be publicly listed
- Firms have to have been listed at least 220 days prior to, and 10 days after, the announcement, in order to estimate wealth effects

This selection makes it possible to capture varying wealth implications due to the synergy communication for the bidder and target firm. Confounding events were also considered in the sample to eliminate non-event-related effects. In the following process, press releases and news reports are researched at the merger announcement of all 60 deals, while recording all communicated, quantified synergy estimates. Therefore, only management estimates were considered, excluding other party estimates – for instance by analysts or journalists. Thereby, the validity, as well as the accuracy, of the data is ensured. Table 3.1 shows the use of voluntary synergy communication over the research timeframe.

Table 3.1: Distribution of M&A transactions by year of announcement

Year	# of Deals	# of Deals with Synergy Disclosure	% of Deals with Synergy Disclosure
1998	7	2	29%
1999	8	4	50%
2000	4	3	75%
2001	4	0	0%
2002	5	3	60%
2003	2	1	50%
2004	2	0	0%
2005	4	2	50%
2006	8	4	50%
2007	6	2	33%
2008	4	4	100%
2009	4	1	25%
2010	2	1	50%
All Years	60	27	45%

Table 3.1 presents the number of transactions and the relative number of deals with disclosed synergies. The sample consists of 60 deals announced between 1 January, 1998 and 31 December, 2010, within the European energy sector.

The synergy communication frequency appears to be stable over time. No indications of a changing willingness to communicate synergy forecasts is found during the period under consideration. For those forecasts lacking precision, the following assumptions were made:

- if a value range is provided, the midpoint is considered;
- if no range, but solely a minimum value, is indicated, the minimum value is considered;

- if no time frame is indicated, the synergies are expected to manifest two years after completion, in accordance with the estimates of most transactions which provide a time frame.

3.5.2 Valuation of Synergy Disclosures

The communicated synergies are discounted to compute their present value (PV) at announcement, following Kaplan and Ruback (1995), while assuming that synergies are realized in perpetuity. The PV is hence computed as follows:

$$PV(Synergies) = \frac{S_t}{(K - I) \times (1 + K)^{t-1}} \tag{3.1}$$

with S being the steady-state communicated synergies after the tth year of completion, K the discount rate, and I the inflation rate. In line with Houston et al. (2001), communicated synergies are expected to grow at the long-term inflation rate at the time of the announcement; thus, zero real growth is assumed. As an inflation estimate, the European harmonized indices of consumer prices (HICP) obtained from the European Central Bank is utilized.

For the calculation of the discount rate K, it is assumed that the communicated synergies carry a risk similar to the bidder's equity, in line with Houston et al. (2001) and Dutordoir et al. (2010). Therefore, K is calculated as:

$$\text{Discount Rate (K)} = r_f + \beta \times \text{market premium} \tag{3.2}$$

with r_f being the risk-free rate, and β the bidder's market beta. As the risk-free rate, the ten-year government bond of Germany is utilized at the time of the merger announcement, following Heston, Rouwenhorst, and Wessels (1999), as well as Croci (2007). The beta is computed utilizing a time period of -40 to -210 trading days prior to the merger announcement, and the S&P Europe 350 as the market proxy. The market premium is assumed to be 7%, in compliance with prior studies on synergy effects (see Houston et al., 2001).

3.5.3 Econometric Design

To measure short-term wealth implications for bidder, as well as target, shareholders, the event study methodology (Fama, Fisher, Michael, & Roll, 1969; MacKinlay, 1997) is applied, with the event being the announcement of the merger or acquisition. While assuming the market model and an estimation period of -40 to -210 days prior to the announcement, the abnormal return (AR) by firm i on day t is computed by subtracting the forecasted market model based expected return E(R) from the actual return R by firm i on day t:

$$AR_{i,t} = R_{i,t} - \alpha_i + \beta_i \times R_{m,t} + \varepsilon_{i,t} = R_{i,t} - E(R_{i,t}) \tag{3.3}$$

while α represents the intercept term at time *i*, β the systematic risk determiner at time *i*, and R_m the return of the market index (S&P Europe 350) at time *t*. Both α and β are calculated using an Ordinary Least Squares (OLS) regression. In the initial model it is expected that the error term will be zero. However, Corhay and Tourani Rad (1996) argue that this is not the case, and expect stock returns to be time dependent. This dependency can be described with Generalized Autoregressive Conditional Heteroskedasticity (GARCH) models, which allow for a non-linear dependency in the residual term. Therefore, as a check for robustness, the event study is also completed using a GARCH (1,1) process which describes the error term as:

$$\varepsilon_{i,t}|\Omega_{i,t-1} \sim N(0, h_{i,t}) \tag{3.4}$$

$\Omega_{i,t-1}$ denotes all information outstanding up to t-1 of firm *i*, while $h_{i,t}$ represents the conditional variance of firm *i*. The variance is conditional on the information set of the period until t-1, which itself is not random, but may be observed. The error term is normally distributed, with a mean of 0 and an assumed variance of:

$$h_{i,t} = \Phi_0 + \Phi_1 \varepsilon_{i,t-1}^2 + \Phi_2 h_{i,t-1}^2 \tag{3.5}$$

Therefore, the conditonal variance consits of a constant, the squared error term in the period t-1, and the conditional variance of the past period. These parameters are calculated using a maximum likelihood estimation. All stock price information for the event study analysis is retrieved from Thomson One Banker SDC using daily total returns (RI).

The cumulative abnormal returns (CAR) and cumulative average abnormal returns (CAAR) for different event periods $[t_1;t_2]$ are calculated as below, with N being the number of transactions. The total event period comprises 11 trading days, with five days prior to the announcement and five after the event date *t = 0*.

$$CAR_{i,t} = \sum_{t=t_1}^{t_2} AR_{i,t} \tag{3.6}$$

$$CAAR_{i,t} = \frac{1}{N} \sum_{t=t_1}^{t_2} CAR_{i,t} \tag{3.7}$$

To test for the statistical significance of the retrieved short-term abnormal returns, the study employs three test statistics. Initially, a simple parametric *t*-test is conducted. Additionally, a cross-sectional test is applied, as proposed by Boehmer, Musumeci, and Poulsen (1991), which controls for a potentially increased standard deviation due to the combination of variance information from the event and estimation window. Furthermore, as prior research provides some evidence that non-parametric *t*-statistics are more powerful that parametric ones (Barber & Lyon, 1996), the statistics are completed with the Wilcoxon rank-sum test.

To infer long-term wealth implications, two different test statistics are employed. The Fama-French-3-Factor model (FF3F) is applied as a common representative of a calendar time approach (Fama & French, 1993) and the buy-and-hold-abnormal-return methodology (Barber & Lyon, 1996) to retrieve long-term event time returns. The FF3F approach is followed since it efficiently eliminates any cross-correlation issues faced by other long-term methodologies, such as the BHAR approach. The FF3F computes the firm's abnormal returns by regressing the firm's excess returns (returns less risk-free rate) against the market's excess returns, the difference in returns of small and big firms, i.e. small minues big (SMB), and the difference in returns of high and low market-to-book values, i.e. high minus low (HML):

$$R_{p,T} - R_{rf,T} = \gamma_p + \delta_p \times (R_{M,T} - R_{rf,T}) + s_p \times SMB + h_p \times HML \tag{3.8}$$

$R_{p,T}$ represents a subsample portfolio of all acquiring firms which disclose a synergy forecast. The sample is continuously changing, while each firm is a member for a duration of *n* months following their announcement; thus, the return $R_{p,T}$ in month *T* indicates the average return of those bidding, European energy firms that were active in the *n* preceding months. Thereby, new acquirers are included, while acquirers that have reached the end of the *n* months holding period are excluded. The average portfolio return is computed from equally-, restricted-, and value-weighted perspectives, with sales volume employed as the weight. The concern that calendar-time approaches weight single-occurring events too high, and therefore also undervalue periods of high activity, is addressed in the following manner. Firstly, portfolios are created in which a single stock's weight is restricted to account for up to 25% of the portfolio (restricted-weighted); for portfolios with three firms of less, the remained is allocated to the market. Secondly, instead of utilizing an OLS regression, a weighted least square (WLS) regression is employed, using the number of firms in a given portfolio as weights. Thereby, increased relevance is allocated to portfolios in times of high activity. In addition, the WLS regression controls for arising heteroskedasticity due to a potential clustering of mergers across time. Monthly returns are utilized at this stage, following Fama (1998), to avoid the influence of the *bad model problem* and to attain more robust statistical influences. Moreover, two time horizons are researched: 24 and 36 months. As the risk-free-rate proxy, the ten-year government bond of Germany is employed; as the market portfolio, which represents a reference portfolio, the S&P Europe 350 Energy index is used. The coefficients γ_p, δ_p, s_p, and h_p are the results of the regression analysis, while the intercept γ_p describes the portfolio's average monthly abnormal return. These factors are tested for statistical significance using the standard *t*-test.

For the second long-term research model, the BHAR methodology (Barber & Lyon, 1996) is used. Abnormal returns of firm *i* for the duration of *T* months are derived by deducting the buy-and-hold return of the reference portfolio $R_{benchmark}$ (i.e. S&P Europe 350 Energy) from the buy-and-hold return of firm *i:*

$$BHAR_i = \prod_{t=1}^{T}(1 + R_{i,t}) - \prod_{t=1}^{T}(1 + R_{benchmark,t}) \quad (3.9)$$

BHARs are computed for 12-, 24-, and 36-month holding periods, while the holding period starts with the announcement of the M&A. BHARs across the sample are calculated as equally weighted averages of individual BHAR. As test statistics, a standard *t*-test and the non-parametric Wilcoxon rank-sum test are applied. Long-term retuns are frequently positively skewed, which leads to a negative bias in the *t*-statistics (Lyon, Barber, & Tsai, 1999). To control for this, the Johnson's skewness-adjusted *t*-test is additionally employed.

3.6 Results and Discussion

3.6.1 Size and Drivers of Synergy Forecasts

Overall, 45% of all European energy-related transactions communicated a synergy estimate within the period under consideration. Bernile and Bauguess (2010) conclude for a different time frame (i.e. 1990 to 2005), and for the US energy market, a synergy communication rate of 31%. Overall, it appears that this particular industry has a relatively high communication rate, which also led Dutordoir et al. (2010) to exclude it from their cross-industry synergy research. Industry-wide, they find that 17% of all deals communicated synergies, based on 2,793 transactions in the US market between 1995 and 2008. Another study by Bernile and Bauguess (2010), using a sample of 3,935 US transactions between 1990 and 2005, conclude a synergy communication rate of 23%.

Regulated industries require multiple approval processes; hence, providing a merger objective, as well as a rationale such as a synergy estimate, is likely to decrease the complexity of the regulatory approval process. Thus, firms operating in the energy sector have an additional incentive to reduce information asymmetries, besides increasing the announcement returns at the point of the M&A. Moreover, industries with objective, measurable synergy potential, such as scale and scope economies, increasingly communicate synergies. Therefore, not only does the energy industry have a relatively high communication rate, but so, for instance, does the chemical industry, with 33%. In contrast, industries with less potential, such as healthcare or durable goods, have a low communication rate, at 18% and 11%, respectively (Bernile & Bauguess, 2010).

To enable further a comparison between the synergies in the European energy industry, their present values are computed, and in a second instance are scaled according to the combined market capitalization of the target and bidder firm. Overall, the average (median) bidder beta of the sample is 0.77 (0.73), while the average (median) discount rate K is 9.57% (9.88%). Apparently, European energy firms are less sensitive than the market (i.e. S&P Europe 350) to market movements, which is suggested by the beta being lower than 1. This fact is, to a great degree, driven by the once highly regulated energy sector market environment. Most energy firms were state-run, and thus were, and still are, subject to a stable cash flow, which reduces the overall firm risk. In addition, the European energy market is characterized by an oligopolistic market structure owing to the ongoing consolidation during the 1980s, which created a few large players with considerable financial strength (Bednarczyk, Schiereck, & Walter, 2010).

Scaling the transactions by the combined bidder and target market capitalization that communicate a synergy estimate results in an average (median) value of 10.11% (6.71%). Prior studies on voluntary synergy disclosure applying an identical scaling lead to similar findings: Houston et al. (2001) conclude 13.1% (9.5%) for the US banking industry, Dutordoir et al. (2010), along with Bernile and Bauguess (2010), obtain 11.7% (6.9%) and even 14% (7.1%), respectively, for a cross-industrial analysis. However, Dutordoir et al. (2010), as well as Bernile and Bauguess (2010), do not, in contrast to the present study, assume synergies to grow at a long-term inflation rate. The present study interprets this finding in the sense that the synergy-to-market capitalization ratios within the European energy industry are below those of the cross-industrial ones of other studies. The absolute synergy figures cannot be expected to cause the low ratio, since the discount rate can be expected to be lower compared to the other studies (due to the low beta) which makes the absolute synergy figures large. Therefore, with respect to the oligopolistic

European energy market and its few large firms, the combined market capitalization appears to be relatively high.

Finally, value drivers of synergies within the European market are identified regarding the fifth hypothesis. I expect that the firm relatedness drives synergies and test these assumptions with an OLS regression (while controlling for heteroskedasticity), and the scaled present values of disclosed synergies as the dependent variable, see Table 3.2 on the following page.

Synergies resulting from cross-border transactions are expected to be below those of national ones due to cultural discrepancies which increase the integration costs (Larsson & Finkelstein, 1999). In particular, cultural differences are likely to lead to misinterpretations of objectives, as well as to interpersonal conflicts (Bruton, Oviatt, & White, 1994). In addition, Kavanagh (2002) demonstrates that most Eastern European power plants require infrastructural upgrades to respond to the most current EU norms and standards. Owing to these additional costs, the profitability of the investment for Western Europe acquirers decreases, while the time to generate synergies increases. As the results in Table 3.2 demonstrate, it is possible to confirm this notion for the European energy sector. The significant negative result proves that cross-border transaction are subject to lower synergy estimates.

A negative result is also expected for the measure *business relatedness*, indicating that unrelated transactions lead to lower synergy forecasts. The present study argues that firms that are engaged in similar operations have an increased potential to transfer knowledge or procedures. In addition, operational aspects such as the reduction of unit costs or the usage of distribution channels can be utilized (Montgomery & Singh, 1987). The evidence shows this impact with a negative slope; however, the mean differences are not statistically significant. Hence apparently, the business relatedness between the firms researched is naturally small, since all firms are active within the same sector (i.e. energy), and thus the study actually assesses the effects of sub-sector relatedness. Therefore, due to the lack of statistical significance, it can be inferred that the various sector characteristics within the energy industry do not affect the synergy generation as a whole.

Table 3.2: Synergy value drivers

Variables	(1) Synergies	(2) Synergies	(3) Synergies
Intercept	0.136***	8.110	-0.020
	(4.409)	(0.980)	(-0.380)
Cross-Border	0.026**	-0.049*	-0.045*
	(-2.575)	(-1.875)	(-1.878)
Business Relatedness	-0.074	-0.022	
	(-1.016)	(-0.775)	
Relative Size	-0.030	-0.006	
	(-0.748)	(-0.204)	
Transaction Value (LOG)		0.073***	0.080***
		(3.136)	(4.000)
Bidder Size (LOG)		-0.034**	-0.036***
		(-2.836)	(-4.355)
Target Size (LOG)		-0.000	
		(-0.018)	
Year		-0.004	
		(-0.981)	
Adjusted R^2	0.118	0.318	0.393
N	27	27	27

Table 3.2 reports the OLS regression for the disclosed synergies which have been discounted to their present value at announcement and scaled by the combined bidder and target market value 60 days prior to the announcement. Bidder and target size is calculated using the log of the respective market value 60 days prior to the announcement. The relative size represents a target-to-bidder sales ratio. Cross-border and business relatedness is measured using a dummy variable, with 0 allocated to national and identical SIC transactions. *T*-statistics are indicated in parentheses, and are robust with respect to heteroskedasticity. ***, **, and * indicate statistical significance at the 1%, 5%, and 10% levels, respectively.

Finally, size relatedness effects are investigated. It is expected, according to Ahuja and Katila (2001), that targets and bidders of similar sizes are able to more efficiently integrate knowledge and other operational aspects. Thus, buyer-managers can more instantly recognize redundant processes, which subsequently leads to a more instant and efficient workforce reduction resulting in increased savings, and hence synergies (Krishnan, Hitt, & Park, 2007). In contrast, Seth (1990) shows that size differences lead to increased synergies, arguing that small buyers can increase their market power, as well as generate economies of scale and scope by acquiring larger targets. From the size ratio (i.e. target sales to bidder sales) it can be inferred that an increase in target size reduces the ratio, all else being equal. Hence, if the slope is negative, synergies increase with a decrease in bidder size, and thus bidder size is negatively correlated to synergies. However, this coefficient is not significant, most likely due to the small sample size. Thus, it is not possible to demonstrate that firm size relatedness influences the magnitude of disclosed synergies.

However, it is possible to show that an increase in the bidder size reduces the magnitude of the communicated synergies, and thus eventually reduces shareholder value creation. Larger bidders apparently generate lower efficiencies, as confirmed by previous research from Moeller et al. (2004). In addition, the evidence suggests that the transaction value increases with synergies, which implies that bidding firms' willingness to pay increases with higher predicted synergies.

Besides controlling for heteroskedasticity while applying the test procedues suggested by White (1980) for the OLS, the utilized variables are checked for multicollinearity and normal distribution. As a check for normallity, the Jarque-Bera test is completed which provides no indication that the residuals are not normal distributed (Jarque & Bera, 1980). Furthermore, the variance inflation factors (VIF) are computed for each parameter, as shown in Table 3.3. This factor estimates the correlation of each variable with all remaining variables, instead of providing a solely pairwise correlation, and thus describes how much of the variance of the estimated coefficients is affected by a correlation with the remaining variables. Typically, for a VIF in excess of 10, serious multicollinearity is evident (Marquardt, 1970; Hair, Anderson, Tatham, & Black, 1995). As the largest VIF for the utilized data is 2.78, there is no any indication of multicollinearity.

Table 3.3: Variance inflation factors

Variables	Cross Border	Business	Relative Size	Transaction Value (LOG)	Bidder Size (LOG)	Target Size (LOG)	Year
VIF	1.38	1.33	1.35	2.78	2.24	2.72	1.31

Table 3.3 shows the variance inflation factors for the variables utilized in the OLS regression from Table 3.2.

3.6.2 Short-Term Stock Price Reactions Following Synergy Disclosure

As expected, bidding firms that do not communicate a synergy estimate are not able to generate positive abnormal returns; this is in accordance with prior studies (Morck et al., 1990; Moeller et al., 2005). Panel B in Table 3.4 demonstrates negative CAARs for all event windows except for the announcement date, as well as the window [0;+1], yet these returns are not statistically significant. Of significance is the period [-5;+5], with a negative CAAR of 2.88%.

Table 3.4: CAAR for bidder European energy firms

	CAAR		Boehmer Test	*t*-Test	Wilcoxon Test	
Event window	Mean	Median	z-score	*t*-value	z-score	N
Panel A: Bidding firms with synergy estimate						
[-5;+5]	1.65%	2.91%	1.171	1.233	-1.538	27
[-1;+1]	2.07%	2.22%	1.898*	1.937*	-2.282**	27
[0;+0]	1.29%	1.18%	1.433	1.490	-1.682*	27
[0;+1]	1.46%	1.69%	1.640	1.570	-2.090**	27
[0;+5]	0.73%	1.26%	0.859	0.670	-0.769	27
Panel B: Bidding firms with no synergy estimate						
[-5;+5]	-2.88%	-1.34%	-2.183**	-2.381**	-1.889*	33
[-1;+1]	-0.52%	-0.38%	-0.395	-0.616	-0.393	33
[0;+0]	0.13%	0.31%	0.620	0.200	-0.935	33
[0;+1]	0.17%	0.24%	0.529	0.211	-0.673	33
[0;+5]	-1.07%	0.21%	-0.567	-1.006	-0.355	33

Table 3.4 presents the mean and median cumulative average abnormal returns of bidding energy firms for various event windows surrounding the M&A announcement using the market model. The estimation period ranges from -40 to -210 days prior to announcement. Panel A reports deals in which managers made a public synergy disclosure. Panel B reports the statistics for the remaining deals. ***, **, and * indicate statistical significance at the 1%, 5%, and 10% levels, respectively.

In contrast, energy firms that do communicate a synergy estimate (Panel A in Table 3.4) are able to generate positive abnormal returns. For the five event windows during the ten days surrounding the announcement, positive abnormal returns are observed. The highest CAAR of 2.07% comes from the window [-1;+1] which is also significant. Moreover, the day of the announcement, as well as the day succeeding it, are significantly positive. For longer time frames, i.e. 10 and 20 days past the announcement, the CAARs decline, yet these are not statistically significant (not shown).

Similarly, M&A that communicate synergies also have a positive effect on target announcement returns. Table 3.5 demonstrates that both panels of target firms (with and without a synergy disclosure) have significant positive announcement returns.

In contrast to the bidding firms' announcement returns, target firms are expected to have a positive CAAR, regardless of a potential synergy communication. As prior research demonstrates, target firms most frequently obtain significant positive announcement returns (Andrade et al., 2001; Brunner, 2004). The CAARs for targets with no synergy communication increase shortly around the event date, and reach their maximum in the event window of [-5;+5] with abnormal returns of 7.79% at the 5% significance interval. During the same interval, targets with an associated synergy communication attain a CAAR of 13.57% at the 1% confidence interval.

Table 3.5: CAAR for target European energy firms

	CAAR		Boehmer Test	t-Test	Wilcoxon Test	
Event window	Mean	Median	z-score	t-value	z-score	N
Panel A: Target firms with synergy estimate						
[-5;+5]	13.57%	10.16%	4.615***	4.345***	-3.700***	27
[-1;+1]	11.72%	7.37%	4.145***	3.926***	-3.868***	27
[0;+0]	10.50%	7.53%	5.261***	4.848***	-4.012***	27
[0;+1]	10.64%	7.48%	4.764***	4.476***	-4.012***	27
[0;+5]	9.47%	7.34%	3.741***	3.707***	-3.315***	27
Panel B: Target firms with no synergy estimate						
[-5;+5]	7.79%	3.61%	2.039**	2.735**	-2.421**	33
[-1;+1]	7.31%	4.34%	2.009**	2.847***	-2.778***	33
[0;+0]	6.19%	3.23%	1.782*	2.386**	-2.743***	33
[0;+1]	5.87%	2.29%	1.762*	2.325**	-2.153**	33
[0;+5]	5.89%	1.69%	1.715*	2.130**	-1.706*	33

Table 3.5 presents the mean and median cumulative average abnormal returns of target energy firms for various event windows surrounding the M&A announcement using the market model. The estimation period ranges from -40 to -210 days prior to announcement. Panel A reports deals in which managers made a public synergy disclosure. Panel B reports the statistics for the remaining deals. ***. **, and * indicate statistical significance at the 1%, 5%, and 10% levels, respectively.

Tables 3.6 and 3.7 carry out the event study for another time, but using the GARCH model assumptions, i.e. the error term is assumed to be conditional on h_p instead of a zero variance assumption. Therefore, the prior findings are checked for their robustness. Overall, the initial findings can be confirmed: bidding firms with associated synergy disclosures reveal positive announcement returns and higher returns than those acquirers that do not communicate any synergy gains. Targets with related synergy expectations also show higher returns relative to their counterparts. Therefore, the prior findings are robust, and can be further utilized in the following analysis.

Table 3.6: CAAR for bidder European energy firms (GARCH adjusted)

	CAAR		Boehmer Test	*t*-Test	Wilcoxon Test	
Event window	Mean	Median	z-score	*t*-value	z-score	N
Panel A: Bidder firms with synergy estimate						
[-5;+5]	1.63%	2.55%	1.281	1.311	-1.369	27
[-1;+1]	2.02%	2.61%	2.030**	2.086**	-2.114**	27
[0;+0]	1.12%	1.29%	1.338	1.404	-1.345	27
[0;+1]	1.53%	1.82%	1.745*	1.718*	-2.282**	27
[0;+5]	0.59%	0.83%	0.712	0.528	-0.625	27
Panel B: Bidder firms with no synergy estimate						
[-5;+5]	-2.32%	-0.61%	-1.638	-1.871*	-1.421	33
[-1;+1]	0.08%	0.51%	0.294	0.088	-0.670	33
[0;+0]	0.27%	0.40%	0.648	0.381	-1.260	33
[0;+1]	0.74%	0.86%	1.188	0.945	-1.510	33
[0;+5]	-0.77%	0.42%	-0.327	-0.745	-0.152	33

***, **, and * indicate statistical significance at the 1%, 5%, and 10% levels, respectively.

Table 3.7: CAAR for target European energy firms (GARCH adjusted)

	CAAR		Boehmer Test	*t*-Test	Wilcoxon Test	
Event window	Mean	Median	z-score	*t*-value	z-score	N
Panel A: Target firms with synergy estimate						
[-5;+5]	14.37%	9.68%	4.687***	4.389***	-3.868***	27
[-1;+1]	12.02%	7.66%	4.165***	3.944***	-3.940***	27
[0;+0]	10.41%	8.49%	5.120***	4.687***	-3.892***	27
[0;+1]	10.78%	7.82%	4.611***	4.240***	-3.892***	27
[0;+5]	9.60%	7.10%	3.688***	3.586***	-3.243***	27
Panel B: Target firms with no synergy estimate						
[-5;+5]	8.40%	4.62%	2.172**	3.099***	-2.939***	33
[-1;+1]	7.72%	6.24%	2.090**	3.046***	-2.939***	33
[0;+0]	6.11%	1.97%	1.751*	2.331**	-2.975***	33
[0;+1]	6.18%	1.85%	1.819*	2.470**	-2.350**	33
[0;+5]	6.11%	1.62%	1.755*	2.260**	-1.957*	33

***, **, and * indicate statistical significance at the 1%, 5%, and 10% levels, respectively.

Recalling hypotheses 1 and 2, the fact that that announcement returns for synergy-disclosing energy firms are more favorable will now be demonstrated. Table 3.8 does this by comparing the mean and median difference for non-disclosing and disclosing firms, as well as by an OLS regression on the CARs, and a synergy disclosure dummy variable. Panel A shows that the mean and median differences for various event windows are significantly lower (except for the [-1;+1] event window for target firms) for non-disclosing firms. Moreover, Panel B shows that synergy disclosure has a significant positive influence on the bidder and combined CARs. Therefore, it is possible to demonstrate, while confirming hypothesis 1, that the communication of synergy estimates has a positive influence on the short-term announcement returns of bidding energy firms. Thus, quantitative synergy forecasts are an efficient instrument to decrease information asymmetries in the European energy market, and convey a value-creating M&A transaction to rational investors. For target firms, on the other hand, only a small amount of evidence is retrieved to confirm hypothesis 2, and the market-implied value creation seems to be mainly attributed to the bidding firm. Nevertheless, transactions as a whole that communicate synergies are again significantly value-creating, which further supports the notion that synergy disclosure leads to a value appreciation.

Table 3.8: Synergy disclosure influence

Panel A: Cumulative average abnormal returns						
	Disclosure sample		Non-disclosure sample		Difference	
	Mean	Median	Mean	Median	Means	Medians
Bidder CAR [-5;+5]	1.65%	2.91%	-2.88%	-1.34%	4.53%**	4.25%***
Bidder CAR [-1;+1]	2.07%	2.22%	-0.52%	-0.38%	2.59%*	2.6%**
Target CAR [-5;+5]	13.57%	10.16%	7.79%	3.61%	5.78%	6.55%*
Target CAR [-1;+1]	11.72%	7.37%	7.31%	4.34%	4.41%	3.03%
Combined CAR [-5;+5]	5.65%	5.23%	-0.91%	-0.59%	6.56%***	5.82%***
Combined CAR [-1;+1]	4.21%	3.34%	0.72%	0.19%	3.49%**	3.15%**
Panel B: Regression analysis						
	Bidder CAR		Target CAR		Combined CAR	
Variables	[-5;+5]	[-1;+1]	[-5;+5]	[-1;+1]	[-5;+5]	[-1;+1]
Intercept	-0.029**	-0.005	0.078***	0.073***	-0.009	0.007
Synergy Disclosure	0.045**	0.026*	0.058	0.044	0.066***	0.035**
R^2	0.098	0.060	0.031	0.021	0.157	0.099
N	60	60	60	60	60	60

Table 3.8, Panel A provides descriptive statistics for bidder, target, as well as combined CARs for the [-5;+5] and [-1;+1] event windows estimated using the market model. It also gives a combined CARs value, weighted by bidder and target market value 60 days prior to announcement. The *disclosed sample* comprises deals in which bidding managers disclose a synergy forecast; non-disclosure deals make up the remaining transactions. The mean (median) significance is tested with the *t*-test (Wilcoxon test). Panel B reports the OLS regression for bidder, target, and combined CARs. Synergy disclosure is a dummy variable attributing 1 to deals with an associated synergy estimate. All *t*-statistics are robust with respect to heteroskedasticity. ***, **, and * indicate statistical significance at the 1%, 5%, and 10% levels, respectively.

Finally, hypothesis 3 is addressed while investigating whether the investor's reaction reflects the synergy estimate. Therefore, whether the released information is perceived as containing incremental new information can also be tested, at least to some extent. Table 3.9 indicates the OLS estimates, controlling for heteroskedasticity, for the bidder, target, and combined CARs of firms with an associated synergy disclosure as the dependent variable and the value of projected synergies as an explanatory one. Projected synergies are again scaled by the respective bidder, target, and combined market capitalization. In addition, other control variables are tested to explain value-creating characteristics of synergy disclosing firms.

Besides, it is acknowledged that the number of bidding European Energy firms amounts to 27 and thus underlying normality assumption shall be tested to confirm the robustness of the found predictors and hence also the accurateness and validity of the retrieved implications. To do so, the residuals of each regression are checked for their normal distribution with the Jarque-Bera test procedure (Jarque & Bera, 1980). The evidence suggests that the residuals are normally distributed for all three models in Table 3.9 below. As an example, the quantile-quantile plot of the residuals from model 3 is illustrated in Figure 3.1 below.

Figure 3.1: Quantile-quantile plot

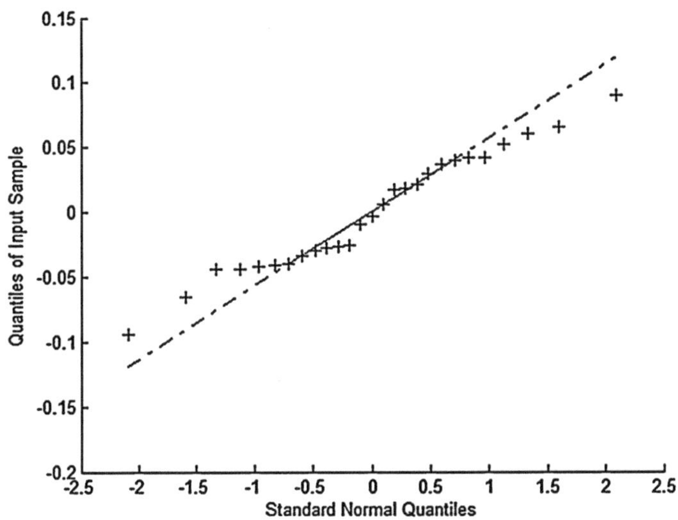

Figure 3.1 displays the distribution of the residuals, indicated by the +, from model 3 in Table 3.9 around a normal distribution.

The evidence of the regression shows that the investor's reaction to the newly formed entity is reflected in the communicated synergies, thus hypothesis 3 is confirmed. Column 3 in Table 3.9 reports a significant, positive relationship between the relative amount of disclosed synergies and the investor's reaction. In particular, a 1% increase in scaled synergies leads to an increase of 36 basis points in the combined bidder and target CAR, all else being equal. Robustness checks yielded qualitatively similar findings for other event windows (not shown); in particular, for the event window [-1;+1] a coefficient of 0.24 results, and for [0], 0.20 is obtained for the combined entity.

Table 3.9: Relationship between CAR and projected synergies

Variables	(1) Bidder CAR	(2) Target CAR	(3) Combined CAR
Intercept	-1.979	10.838	-3.634
	(-0.292)	(0.643)	(-0.630)
Synergy/Bidder Market Value	0.106		
	(1.473)		
Synergy/Target Market Value		-0.052	
		(-1.552)	
Synergy/Combined Market Value			0.363***
			(3.172)
Bidder Size (LOG)	-0.009	0.053	-0.026**
	(-0.822)	(0.977)	(-2.110)
Target Size (LOG)	0.005	-0.065	0.009
	(0.527)	(-0.970)	(0.857)
Relative Size	-0.068***	-0.109	-0.043**
	(-2.819)	(-1.179)	(-1.970)
Cross Border	0.046**	0.003	0.015
	(2.200)	(0.058)	(0.745)
Business Relatedness	0.007	-0.023	-0.005
	(0.343)	(-0.361)	(-0.249)
Year	0.001	-0.005	0.002
	(0.296)	(-0.629)	(0.651)
Adjusted R^2	0.042	-0.167	0.219
N	27	27	27

Table 3.9 reports the OLS regression on the bidder, target, and combined CARs for the [-4;+1] event window, following Houston et al. (2001), of those transactions with an associated synergy disclosure. It also provides a combined CARs value weighted by bidder and target market value 60 days prior to announcement. Present value synergies are scaled by the respective market value 60 days prior to the announcement. Bidder and target size is calculated with the log of the respective market value 60 days prior to the announcement. The relative size represents a target-to-bidder sales ratio. Cross-border and business relatedness is measured with a dummy variable allocating 0 to national and identical SIC transactions. The *t*-statistics are indicated in parentheses and are robust with respect to heteroskedasticity. ***, **, and * indicate statistical significance at the 1%, 5%, and 10% levels, respectively.

Moreover, checks were conducted for the synergy discount rate assumptions in Section 3.6.3. Columns 1 and 2 in Table 3.9 show that the synergy size does not affect the individual CAR of the bidder and target. Concerning target firms, this finding is not surprising, since the respective CARs were not found to increase upon a synergy disclosure (see Table 3.8), and thus the synergy magnitude should also not affect the abnormal returns. Moreover, since bidding CARs do not correlate to the magnitude of synergies, the market apparently values synergies primarily with respect to the transaction as a whole. Nevertheless, the existence of synergies is appreciated, and therefore reflected in the bidder CARs, as shown in Table 3.8.

Nonetheless, if the stock market fully accounts for the amount of disclosed synergies as incremental information, the slope should be 1; however, this is not the case, as the slope is 0.36. Therefore, one interpretation is that the market discounts the management's synergy projection. Hence, the market assumes that European energy managers are likely to overestimate the actual, realizable synergies (Fischer & Verrchecchia, 2000). With reference to the US banking sector, the market discounts projected synergies to a lesser extent, with a slope of 0.49 (Houston et al., 2001). An explanation for this may be that managers and their estimates in the European energy sector are perceived as being less credible. Nonetheless, when the results are compared to an industry average reaction to synergy disclosure, managers in the energy sector are attributed with average credibility. The industry average, which is solely calculated for the [-1;+1] window, is, according to Bernile and Bauguess (2010), 25 basis points; the respective slope for the European energy sector is 0.24. On the other hand, as Houston et al. (2001) suggest, the markets frequently anticipate the M&A, and thus at announcement potential gains are already partially priced in the stock, which consequently reduces the respective bidder and target reaction.

3.6.3 Robustness Tests

In the following, the assumptions on the present value computation of the disclosed synergies are tested for robustness. This not only validates the chosen model, but also investigates the effects of certain deviations. The model itself is based on two key assumptions, i.e. the market premium, which is assumed to be 7%, and the synergy realization horizon, which is expected to be perpetual. Thus, the focus is placed on these parameters.

Initially, the above-described regression is completed with a changed risk premium. The premium is deviated by 0.5% to 6.5% and 7.5%, respectively columns 1 and 2 in Table 3.10. The evidence shows shows that the significant postive influence of the synergy size on the abnormal returns holds. Hence, the implication that the market is able to price the magnitue of disclosed synergies is confirmed, even though the coefficients of this parameter deviate slightly.

Table 3.10: Market premium (robustness)

Variables	(1) 6.5% - Premium Model	(2) 7.5% - Premium Model
Intercept	-3.155	-3.428
	(-0.513)	(-0.564)
Synergy/Combined Market Value	0.262**	0.300**
	(2.249)	(2.300)
Bidder Size (LOG)	-0.026*	-0.025*
	(-2.038)	(-2.017)
Target Size (LOG)	0.009	0.009
	(0.835)	(0.841)
Relative Size	-0.036	-0.036
	(-1.466)	(-1.440)
Cross Border	0.012	0.012
	(0.632)	(0.639)
Business Relatedness	-0.011	-0.010
	(-0.463)	(-0.447)
Year	0.002	0.002
	(0.533)	(0.584)
Adjusted R^2	0.150	0.154
N	27	27

Table 3.10 reports the OLS regression on the combined CARs for the [-1;+4] event window and those transactions with an associated synergy disclosure. Models (1) and (2) deviate from the initial market premium assumption while subtracting and adding 0.5%, respectively. Combined CARs are computed by allocating equal weights to bidder and target CARs. Present value synergies are scaled by the respective market value 60 days prior to the announcement. Bidder and target size is calculated with the log of the respective market value 60 days prior to the announcement. The relative size represents a target to bidder sales ratio. Cross-border and business relatedness is measured with a dummy variable allocating 0 to national and identical SIC transactions. *T*-statistics are indicated in parentheses, and are robust with regards to heteroskedasticity. ***, **, and * indicate statistical significance at the 1%, 5%, and 10% levels, respectively.

In a second instance the initial assumption (which is in accordance with all prior synergy-related research) that synergies realize in perpetuity is revised to the assumption that synergies realize solely up to five, seven, and ten years following the M&A completion. The results remain qualitatively the same, yet with a distinctive difference. It can still be concluded that the market values the synergy size, as the coefficient is significantly positive at the 5% interval; however, the coefficient increases from 0.36 for the perpetuitiy assumption, to 0.86 for the seven-year model. From this, it can be concluded that the market discounts disclosed synergies only by a small margin if the synergy horizon is smaller, and thus believes that the combined entity is able to implement a higher proportion of the communicated value generation. Apparently, the further away the expected integration horizon is, the lesser believes the market in its full realization.

Table 3.11: Synergy realization horizon (robustness)

Variables	(1) 5-year Model	(2) 7-year Model	(3) 10-year Model
Intercept	-7.586	-6.721	-5.568
	(-1.307)	(-1.177)	(-0.970)
Synergy/Combined Market Value	1.041***	0.857***	0.659**
	(3.453)	(3.078)	(2.778)
Bidder Size (LOG)	-0.007	-0.012	-0.016
	(-0.729)	(-1.211)	(-1.601)
Target Size (LOG)	0.008	0.007	0.007
	(0.837)	(0.714)	(0.674)
Relative Size	-0.427	-0.040	-0.037
	(-1.673)	(-1.604)	(-1.538)
Cross Border	0.019	0.018	0.017
	(1.053)	(1.006)	(0.899)
Business Relatedness	0.006	0.002	-0.002
	(0.277)	(0.109)	(-0.067)
Year	0.004	0.003	0.003
	(1.315)	(1.189)	(0.986)
Adjusted R^2	0.285	0.267	0.236
N	27	27	27

Table 3.11 reports the OLS regression on the combined CARs for the [-1;+4] event window, and those transactions with an associated synergy disclosure. Models (1), (2), and (3) assume a synergy realization horizon of 5, 7 and 10 years, respectively. Combined CARs are computed by allocating equal weights to bidder and target CARs. Present value synergies are scaled by the respective market value 60 days prior to the announcement. Bidder and target size is calculated with the log of the respective market value 60 days prior to the announcement. The relative size represents a target-to-bidder sales ratio. Cross-border and business relatedness is measured with a dummy variable allocating 0 to national and identical SIC transactions. *T*-statistics are indicated in parentheses and are robust with regards to heteroskedasticity. ***, **, and * indicate statistical significance at the 1%, 5%, and 10% levels, respectively.

Therefore, if the realization horizon is further extended up to ten years, the coefficient remains significantly positive at the 5% interval, yet the magnitude again decreases to 0.66. This finding can be related to the unique industry environment of the European energy market. As the market is still in a state of transition, with ongoing consolidations and frequent regulatory changes, a long-term operational projection is difficult to make. Thus, foreseeing operational processes such as the integration of synergies from a today's perspective is problematic, and as a precautionary measure capital markets apparently discount more distant synergies more strongly.

3.6.4 Long-Term Stock Price Reactions Following Synergy Disclosure

The following section explores whether bidding firms can sustain their significant positive announcement returns utilizing the FF3F model, and hence also addresses hypothesis 4. Table 3.12 shows the intercept of the FF3F model, indicating the average monthly abnormal returns for the two- and three-year holding period of synergy-disclosing energy firms.

The results demonstrate significant positive returns for both holding periods; thus, synergy-disclosing deals experience abnormally high stock returns. In particular, positive monthly average abnormal returns ranging from 0.7% to 1.7% are retrieved for the initial two years following the M&A; hence, for the respective holding period, there is a compounded value appreciation of 16.8% to 40.8%, respectively. These findings are robust against a change in portfolio weighting, and moreover can be confirmed by the three-year holding period. Thus, there is evidence that synergy-disclosing energy firms can sustain their short-term positive abnormal returns over a three-year time horizon. As Fama and French (1997) demonstrate, positive abnormal returns are usually associated with a reduction of risk loadings. Therefore, the voluntary synergy disclosure and thus reduced information asymmetries are perceived as a risk reduction by the market. I interpret this finding in a way that synergy-disclosing mergers become unexpectedly more profitable, and hence, from an investor's point of view, less risky. Similar findings are encountered for comparable information-asymmetry-reducing efforts such as dividend announcements. As such, Boehme and Sorescu (2002) report average monthly abnormal returns ranging from 0.77% to 1.41%.

Table 3.12: Abnormal returns to bidding firms based on the FF3F model

	Horizon = 2 years (No. Obs. = 148)		Horizon = 3 years (No. Obs. = 148)	
Calendar Portfolio Weighting	Intercept of Fama-French Regression	Adjusted R^2	Intercept of Fama-French Regression	Adjusted R^2
Equally-weighted	0.017***	0.606	0.013**	0.366
	(3.083)		(2.022)	
Restricted-weighted	0.007***	0.833	0.012**	0.444
	(2.162)		(2.143)	
Value-weighted	0.017***	0.603	0.013*	0.347
	(2.795)		(1.931)	

Table 3.12 reports calendar-time abnormal returns for synergy-disclosing European energy firms using the FF3F model. Portfolio returns are reported equally-, restricted-, and value-weighted. In the restricted version, single-stock weightings are limited to a maximum of 25%. For the value-weighted version, sales are utilized as weights. All statistics were performed with a WLS regression; *t*-statistics are indicated in parentheses. ***, **, and * indicate statistical significance at the 1%, 5%, and 10% levels, respectively.

These findings thus suggest that synergy-communicating firms outperform their market environment in the years immediately following the merger. Yet, the longer the holding period, and thus the period both entities are integrated, the more the excess return declines; the monthly abnormal return declines at about 0.4%. This finding is interpreted in the sense that the capital market is partially not confident that the management is able to sustain and continuously generate the communicated synergies once integrated. For instance, structural reorganization or cost-cutting efforts which initially create value might, over time, become more lax, and thus incremental merger benefits partially diminish.Consequently, abnormal returns should also diminish, as demonstrated here.

Alternatively, competitive market pressure could lead to an overall increase in efficiency of all market participants, which hence decreases the relative higher performance of synergy-disclosing M&A transactions. Therefore, the performance gap converges over time. In the context of the European energy market, this increase in market pressure is likely to manifest as all ongoing consolidation efforts are likely to also increase firm efficiency, which eventually increases the consumer welfare. Nevertheless, it is possible to demonstrate that the management estimates considered here were not irrational or based on management overconfidence, and support the view already established for the short-term effects, that synergy disclosure leads to value creation, particularly for the bidding European energy firm.

The long-term analysis utilizing the BHAR methodology concludes qualitatively similar findings. As Table 3.13 illustrates, significantly positive returns are retrieved for the two- and three-year holding period. Their magnitude of 16% and 30%, respectively, can be regarded as fairly consistent with those of the FF3F model, supporting the conclusion that synergy-disclosing firms in the energy sector are able to implement and realize expected synergies.

Table 3.13: Abnormal returns to bidding firms based on buy-and-hold-abnormal-returns

Holding Period	BHAR		*t*-Test	Johnson Test	Wilcoxon Test	No. Obs.
Months	Mean	Median	*t*-value	J-value	z-score	No.
12	5.99%	4.95%	1.111	1.124	-0.978	26
24	16.18%	8.63%	1.781*	1.854*	-1.673*	23
36	30.64%	7.28%	2.388**	2.539**	-1.790*	21

Table 3.13 reports equally weighted buy-and-hold-abnormal-returns (BHAR) for the duration of 12, 24, and 36 months for acquiring, synergy-disclosing European energy firms. ***, **, and * indicate statistical significance at the 1%, 5%, and 10% levels, respectively.

3.7 Implications for European Energy Firms

As the final section in the chapter, value-increasing recommendations are developed for energy firms pursuing M&A within the EU. These are particularly designed with regards to the present challenges managers face. Thereby, those transactions which lead to the highest synergy potential are identified, as well as the respective communication strategy.

Product bundles, especially convergent mergers between electricity and gas, are a frequent objective in the European M&A market. While studies focusing on the US market (Studness, 1996) show a value destruction from these mergers, the present findings do not identify any value-decreasing consequences; there is, however, also no additional synergy potential to be gained by striving for convergent or focused M&A. Hence, the results imply no value loss for bundles, and as there may be other value-creating aspects relating to diversifying M&A, such as the reduction of firm risk by establishing two unrelated lines of business, convergent mergers eventually result in a wealth appreciation. In addition, the results show that the geographic M&A strategy may have a significant impact within the European energy market. A lower synergy potential is found for cross-border mergers between energy firms, and, as the synergy size is reflected in the capital market reaction, cross-border M&A lead to lower announcement returns in contrast to national mergers. Therefore, small firms in particular, that require a positive market valuation, should firstly pursue national transactions to not only extract the maximum synergy potential, but also increase the probability of actually completing the deal, which is highly correlated to the announcement returns as shown by Luo (2005).

The preceding study shows increased returns for synergy-disclosing mergers, yet also demonstrates that markets discount these synergies. Hence, if the transaction is intended to lead to synergetic gains, it is strongly recommended that these be disclosed to the capital market at the announcement. Potential expenses related to measuring synergy gains *ex ante* can reasonably be expected to be offset due to positive announcement returns. While only 45% of the firms in the sample disclose synergies, the remaining transactions can partially also be expected to lead to synergies; as shown here, business sectors in the energy environment and firm size differences are not distinctly different, nor do they represent synergy-reducing elements. Thus, these firms eventually prevent the capital market from perceiving the value of their merger by failing to disclose synergies. Nevertheless, energy managers and their synergy projections are not perceived to be fully credible by the capital market, as these synergy estimates are discounted. It is likely that the driving force is the lack of detail, and therefore the remaining uncertainties. Prior research by Kimbrought and Louis (2011) shows a positive causality between conference call details and M&A announcement returns. Therefore, as synergy disclosures and conference calls exhibit similar characteristics, synergy-disclosing energy firms should not solely disclose pure annual synergy figures (as most of the firms in the sample do), but should enrich these by adding further details. Aspects that should be considered include:

I. where synergy gains result from;
II. what their growth rates are in the initial years proceeding the completion; and
III. which actionable measure are taken to realize disclosed synergies.

Particular attention needs be placed on the regulatory environment of the EU energy market. An analysis regarding the degree to which the realization depends on the current market environment, and how regulatory changes may affect the synergy realization, is recommended.

3.8 Conclusions

This third chapter investigated the synergies and their drivers in the European energy industry, as well as the effects of their disclosure on the respective bidder and target announcement returns.

The evidence suggests that synergy disclosure and size within the European energy sector significantly differs from other industries. Not only do bidder-managers communicate synergies more frequently, with 45% of all transactions, but the relative size of those communicated synergies are also smaller, compared to the industry average. These findings can be redirected to the unique industry characteristics, with the European energy sector being highly consolidated and thus characterized by an oligopolistic market structure. The investigation of synergy drivers proves that the relatedness of the M&A affects the disclosed magnitude. In particular, the cultural component revealed its significance while other parameters tested (i.e. size and business) were not found to shape synergies significantly. Thus, cross-sectional transactions within the energy market between various utilities apparently do not affect the synergy potential.

The findings of the voluntary synergy disclosure on announcement returns are consistent with other studies on comparable aspects (Kimbrought & Louis, 2011). Hence, synergies serve as an instrument to reduce information asymmetries, and may also be utilized as a value-signaling tool. Target and bidder firms (as well as the combined entities) which disclose quantitative synergy forecasts experience excess announcement returns, in contrast to firms which do not do this. Moreover, the magnitude (i.e. the amount of synergies scaled by the combined target and bidder market capitalization) directly relates to the magnitude of abnormal returns. The study demonstrates that a 1% increase in scaled synergies leads to an increase of 36 basis points in the combined bidder and target CAR. This relationship also provides evidence that the market discounts the disclosed synergies, and hence the market believes that managers are likely to overestimate the actual, realizable synergies to a certain degree. Nonetheless, in the long run these firms are able to sustain their positive announcement returns and continuously create value for their shareholders, as over a two-year holding period bidders experience average monthly abnormal returns ranging from 0.7% to 1.4%, depending on the methodology.

A fruitful further area of research emerging from the study is an investigation of determinants of management credibility in disclosing synergy forecasts. An understanding of these factors will enable managers to further increase their respective announcement returns. Other relevant aspects are varying market perceptions of different synergy sources, i.e. an investigation of announcement returns by communicating financial, organizational, or revenue synergies.

4 The Interaction of Synergy Disclosures and Post-Merger Operating Performance[3]

4.1 Introduction

One of the most fundamental organizational changes are M&A; however, the consequences of such unique events are difficult to foresee. Voluntary information disclosures are an often-used instrument to help investors to estimate the future prospects of their security position. Thus far, the research focus has been on purely accounting-related supplementary disclosures prior to M&A, such as earning forecasts (Erickson & Wang, 1999; Louis, 2004), disclosure frequency (Lang & Lundholm, 2000; Jo & Kim, 2007), or conference calls (Bushee et al., 2004; Kimbrought & Louis, 2011). More recently, the voluntary disclosure of emerging synergy effects has gained substantial importance as they have been proven to significantly increase the value of the transaction, as the prior chapter reveals. Preceding studies show that bidding firms frequently experience value destruction at the M&A announcement (Morck et al., 1990; Moeller et al., 2005) owing to existing information asymmetries between the bidding firm and market (Moeller et al., 2007; Officer et al., 2009). Yet, if quantitative synergy estimates are disclosed along with the announcement, investors are not only informed about the merger motive, but also about the implied wealth gain, which subsequently serves as a value-creation signal for rational market participants. The resulting reduction of information asymmetries thus increases merger-implied gains. As such, synergy-disclosing transactions have been proven to increase announcement returns for bidding firms across industries (Bernile & Bauguess, 2010; Dutordoir et al., 2010).

On the contrary, the question of whether these disclosed synergies are rationally justified and, hence, operationally realizable, nevertheless has to be asked, particularly as the first chapter of this dissertation shows that the disclosure of synergies is also utilized as a political instrument. The following research responds to this issue by assessing the post-merger operating performance of synergy-disclosing transactions against non-disclosing deals, as well as non-merging firms. In this way, the capital-market-driven view of voluntary synergy disclosures is extended by the operational counter-perspective, and a demonstration is given as to whether excess announcement returns are based on management overconfidence, market overreaction, or operational reasonability. By doing so, the honesty in managerial reporting, which has proven by Evans

3 I am grateful to Prof. Dr. Dirk Schiereck for helpful insights and comments.

III, Hannan, Krishnan, and Moser (2001) along with Matuszewski (2010), to be a substantial value-creating factor, can, to a certain degree, also be assessed for synergy disclosures.

In efficient markets, new information, such as the announcement of M&A, is rational and immediately added to a firm's value. Therefore, merger-related efficiency gains (i.e. synergies) are incorporated immediately. Houston et al. (2001) show this in their research on the US banking sector using a short-term event study, and moreover demonstrate that the firm's market value increases with increased, expected efficiency gains, yet only if these are disclosed to investors. The present study is able to prove a similar effect for the European energy sector. On the other hand, this implies that deals which do not disclose synergies experience a poorer capital market valuation at announcement, as the market does not attribute these particular transactions to be also intended to realize any synergies in the post-merger horizon. However, if long-term event studies are analyzed, the market adjusts for this short-term abnormal positive reaction of the synergy-disclosing deals, according to Bernile and Bauguess (2010), and therefore these transactions experience a significantly negative long-term performance. A possible interpretation of this is that the market was overconfident that the M&A would actually realize the disclosed efficiency gains. Hence, from the market's perspective, expected synergies did not materialize and therefore short-term excess returns were not operationally justified, as disclosing deals were not able to outperform non-disclosing ones.

However, even though this conclusion is tempting, in such a long-term horizon various value-influencing and potentially value-reducing events could have occurred, and thus interpreting the value loss as not being able to realize disclosed synergies might not necessarily be the case. As such, Dutordoir et al. (2010) are not able to show that synergy-disclosing deals have a poorer capital market long-term performance; this directly contradicts Bernile and Bauguess (2010) as Dutordoir et al. (2010) show that disclosing deals exhibit a significantly better capital market performance. Similarly, the preceding chapter also indicates that synergy-disclosing European energy firms experience an increased long-term market valuation. However, the applied methodology, i.e. buy-and-hold abnormal return, and respective model assumptions are frequently critiqued (see Fama & French, 1993). Thus, overall, long-term event studies do not provide conclusive evidence of whether short-term excess returns are justified by expected synergies, particularly as existing studies consider a cross-industrial perspective with multiple value-influencing events that are not controlled for.

Therefore, the present study now uses an operationally driven approach to conclude upon the realization of synergies and thereby excludes the market's valuation from the analysis. To the author's knowledge, no prior operational performance study has distinguished M&A according to the merger motive, and thus whether the deal objective is to actually generate synergy gains. Prior research treats all deals equally while testing all M&A for an improvement in performance, yet it may be that only some transactions are value-creating, and, on average, statistics are canceled out (as transactions with different motives are eventually combined). To capture this synergy effect, this study looked for an industry that is characterized by a high ratio of voluntary synergy disclosure and similar business models between its sectors. The energy industry was therefore selected, as it has one of the highest synergy disclosure rates, at 30.7%, in the US (Dutordoir et al., 2010). To further increase sample homogeneity with respect to general market conditions, the geographic focus is on the US energy sector.

Few studies analyze the operational performance of the energy sector, while the majority researches the influence of market deregulation. As such, Ramos-Real, Tovar, Iootty, Fagundes de Almeida, and Queiroz Pinto Jr. (2009) and Sueyoshi and Goto (2011) show that deregulation decreases efficiency, while Cullmann and Hirschhausen (2008) demonstrate an increase in efficiency due to deregulatory efforts, with the evidence again being inconclusive. With respect to M&A, Kwoka and Pollitt (2010) conclude for the US market that M&A do not lead to an increase in efficiency for both the bidding and target firm. Becker-Blease, Goldberg, and Kaen (2008) further support the notion that M&A-conducting utility firms are unable to outperform their counterparts, and thus propose that if synergies evolve within the deregulated market environment, M&A-inducing utilities are not able to capture these. However, these findings directly contradict event-study-implied value gains, as these studies unanimously show a value generation for all target firms, and moreover for bidding firms with associated synergy disclosures. Assuming the merger motive is announced along with the M&A, only a fraction of all transactions in the energy sector actually generate synergies (Bernile & Bauguess, 2010). Therefore, the implication that M&A in the energy sector are unable to generate or capture synergies could be wrong; by examining the actual, announced merger motive and testing only those deals which intend to generate efficiency gains, it is expected that the present study will be able to retrieve these. The remaining transactions might pursue varied objectives such as increasing market power by creating monopolies, creating market discipline by switching a target's management team, pursuing opportunities of diversification, or being driven by managerial hubris, and thus overall do not exhibit efficiency gains (Andrade et al., 2001).

This proposition gains in relevance as event studies which specifically focus on announced merger objectives find significant differences for disclosing and non-disclosing transactions. As such, Houston et al. (2001), Dutordoir et al.

(2010), and Bernile and Bauguess (2010) demonstrate that M&A which announce cost reductions are perceived by the market to create additional value, in contrast to those transaction that do not do so. Therefore, the announcement of the merger objective and, specifically, the announcement that the deal is expected to enable synergies has a distinctive influence on the value perception. In the following section, this notion is transferred to operational efficiency studies. Thus, the section is intended to show that M&A can, in contrast to prior research, result in efficiency gains if the transaction is structured to do so. In a subsequent analysis, these firms may then serve as a peer group, and, by estimating the efficiency drivers, other transactions are enabled to also explore their potential to create value.

The remainder of this chapter is structured as follows. Section 4.2 conducts a literature review. Section 4.3 identifies the research model, while section 4.4 presents and analyzes the findings. Section 4.5 develops implications for M&A-seeking managers. Section 4.6 concludes.

4.2 Literature Review

Three motives induce managers to conduct an M&A: market power, hubris, and synergies (Berkovitch & Narayanan, 1993). The latter proposes that mergers are a method to enhance efficiency, in particular to increase the performance of their assets, or to utilize existing assets which were previously underperforming, in such a manner that these create additional value. In order to identify these merger-related gains, two primary streams of research have evolved: capital-market-oriented event studies and accounting-related research.

Overall, as Andrade et al. (2001) show while summarizing the vast literature upon this research instrument, event studies result in value creation for the combined entity, while the majority of the generated wealth is transferred to the target firm. This implies that targets usually experience positive announcement returns (Moeller et al., 2004), while acquirers frequently (Morck et al., 1990; Moeller et al., 2005) gain either zero or negative returns. In the industry of interest, the energy sector, these general results can be confirmed. As such, Leggio and Lien (2000) show, with a sample of 76 electricity M&A between 1983 and 1996 in the US, that targets experience significantly positive returns, while the bidding entity's share price decreases significantly. Berry (2000) also confirms these results, and furthermore shows that the combined firm within the energy sector leads to value creation. However, in the long-term, as Gugler, Mueller, Yurtoglu, and Zulehner (2003) show while analyzing five-year merger performance, the majority of M&A transactions are welfare-reducing.

On the other hand, accounting-related research provides an inconclusive perception of M&A. Most notable, Healy, Palepu, and Ruback (1992) find (while utilizing the intercept model) that in the post-merger period bidder cash flows increase significantly. Schwitzer (1996) is able to confirm this notion. Yet, their research uses an industry median portfolio as the benchmark, which is likely to bias the findings as M&A-conducting bidders have been proven to exhibit significantly different pre-merger characteristics than the market (Morck et al., 1990). Therefore, Gosh (2001) duplicates this analysis with a pre-merger performance-matched benchmark, and is thereby not able to detect any abnormal increase in post-merger operating cash flows. On the contrary, Linn and Switzer (2001), while also utilizing a pre-merger performance-matched benchmark, find a statistically significant increase. Applying a different methodology, the change model, research by Gosh (2001) is yet still not able to prove any performance increase following the merger. As an attempt to solve this discrepancy, Powell and Stark (2005) conduct a comprehensive study upon methodologies and benchmark indices. They find that while the intercept model leads to higher post-merger performance increases, the chosen benchmark and its composition does not influence the outcome. Moreover, they demonstrate that M&A lead to value creation, and the method of payment does not influence the respective magnitude.

For the industry of interest, the energy sector, a few methodically, as well as conceptually, similar studies have been conducted, again showing varied value perceptions. The effects of energy market deregulation on efficiency gains has been analyzed by Sueyoshi et al. (2009) and Sueyoshi and Goto (2011) for the US market. Thereby, they distinguish between financial, corporate value, and operational synergies. Utilizing data envelopment analysis (DEA), they show that within deregulated markets, financial as well as corporate-value-related synergies diminish if energy firms follow a diversification strategy (Sueyoshi et al., 2009). A proceeding study analyzing operational synergies, again utilizing DEA, leads to similar conclusions. In the post-deregulation period, corporate diversification does not result in efficiency gains (Sueyoshi & Goto, 2011). Similar results are retrieved for the Brazilian energy market, as reforms do not lead to improved efficiency (Ramos-Real et al., 2009). On the other hand, Cullmann and Hirschhausen (2008) illustrate that the deregulatory efforts in the Polish energy sector have value-increasing consequences. They show, for 32 energy firms during 1997 and 2002, an increase in technical efficiency, as well as increasing returns to scale. Therefore, a market consolidation through M&A would further increase the efficiency exploration. Overall, the evidence of efficiency gains due to the market deregulation is mixed, and apparently depends on the deregulatory policies. Nevertheless, none of these studies provide evidence on the fundamental question of whether synergy-expecting M&A lead to an improvement in efficiency.

Only a few studies have directly addressed the influence of M&A on efficiency. Becker-Blease et al. (2008) show that M&A in the energy industry do not result in increased efficiency. Analysis of M&A between 1992 and 2001 within the US shows that M&A-conducting, relative to non-M&A-conducting, energy firms yield lower profitability measures in the post-merger period. Similar results are retrieved by Kwoka and Pollitt (2010) for the US market. Here, bidding, as well as target, firms were not able to improve their efficiency following M&A, and moreover target firms' efficiency scores even declined. On the contrary, Hattori, Jamasb, and Pollitt (2005) research the comparative efficiency of electricity distributions systems in the UK and Japan during 1985 and 1998. Overall, they find that UK-based firms experienced a greater productivity gain than Japanese ones during 1994 and 1998. This period was characterized by an increased rate of M&A, which hence fostered an improved operation efficiency. They thus support the notion that mergers lead to value creation. Moreover, Kwoka (2005) shows that mergers between electricity firms have the potential to generate significant economies of scale within major operational functions (i.e. distribution, wires, and supply), and hence generate synergies. Furthermore, Fabrizio, Rose, and Wolfram (2007) demonstrate that restructuring and ownership-change leads to substantial medium-run efficiency gains. On the whole, the evidence is again mixed, and the current debate concludes that M&A in the energy sector have efficiency potential, though contemporary studies have not been able to confirm this.

The research in the following section builds on these prior studies, yet with distinctive differentiations. First, not all M&A are considered to be equal, but the M&A motive (i.e. a synergy disclosure) is specifically considered, making it possible to investigate the operational performance in more depth, as well as show that synergy disclosures are a signal for an increase in the post-merger operating performance. Second, the sample size of completed M&A is extended by 50% relative to Becker-Blease et al. (2008), in order to infer more robust conclusions. Finally, multiple regression analysis is conducted to identify those characteristics which foster the performance in- or decrease.

4.3 Data and Methodology

4.3.1 Data

The Thomson One Banker SDC database is employed to retrieve the relevant information for all energy-related M&A occurring between 1 January, 1997 and

31 December, 2008. To identify the respective transactions, the following selection criteria are applied:

- Both the bidding and target firm have a primary SIC code of 1311 (Crude oil and natural gas), 4911 (electric services), 4922-25 (natural gas transition and distribution), 4931-32 (electric/gas and other services), or 4941 (water supply).
- Both the bidding and target company were publicly listed for at least 273 trading days prior to, and 10 days following, the transaction, which gives an estimation period of one trading year (i.e. -20 to -273).
- Financial statements for two fiscal years prior to and following the transaction are available on the Thomson One Banker Company Analysis Database.
- The transaction has a minimum deal value of US$ 1m.
- The transaction is complete and a voting stake (min. 50% transfer of voting shares) is acquired.

The focus is deliberately on the target to be publicly traded in order to investigate diverging wealth effects of a synergy disclosure on the respective ownership entity. In addition, operational performance analysis is conducted for the period of three years prior to and following the merger; however, owing to limited data availability, the sample size decreases by six non-disclosing M&A for these time periods. Furthermore, following Morck et al. (1990), a minimum transaction value is required in order to exclude small transactions, as these have been proven to bias the measurement of performance. Stock market data is retrieved on a daily basis, and has been adjusted for any dividend payments (i.e. total return index (RI)). Furthermore, confounding events are looked for in the sample in order to eliminate non-event-related effects. The final database consists of 80 M&A transactions, which are then researched for a respective synergy disclosure at the announcement. The Thomson First Call Database is not used therefore, as it has proven to be incomplete (Chuk, Matsumoto, & Miller, 2012); instead, the synergy disclosures were hand-collected. Thereby, only quantitative, management-released forecasts are considered to ensure their credibility and accuracy; analyst or newspaper estimates are excluded. Finally, 25 transactions (i.e. 45%) disclose synergy forecasts; Table 4.1 shows the distribution of deals per year.

Table 4.2 provides information about synergy-disclosing and non-disclosing bidder characteristics, respectively. Balance sheet items are taken from the end of the fiscal year preceding the announcement.

The analysis highlights that disclosing firms distinctively differ from their counterparts in terms of absolute size. Pure size measures are significantly higher for synergy-disclosing firms, yet corresponding ratios appear to be constant across both subsamples. Thus, the capital structure indicated by the debt-to-assets ratio, as well as the firm performance of the assets, is comparable across the sample.

Table 4.1: Distribution of M&A transactions by year of announcement

Year	Overall	# of Related Deals	# of Synergy-Disclosing Deals
1997	14	10	2
1998	7	3	3
1999	16	7	8
2000	8	5	2
2001	7	5	2
2002	2	2	0
2003	3	2	2
2004	5	4	1
2005	6	5	3
2006	6	4	1
2007	4	2	1
2008	2	2	0
Total	80	55	25

Table 4.1 presents the distribution of deals per year in the US energy sector which were announced between 1 January, 1997 and 31 December, 2008. Transactions are labeled as related if their SIC is identical.

Table 4.2: Descriptive statistics of acquiring firms

	Acquirer without Synergy Disclosure		Acquirer with Synergy Disclosure		Difference	
	Mean	Median	Mean	Median	Mean	Median
Sales ($mil)	3,206.70	1,083.90	18,939.68	5,346.43	15,732.97**	4,262.53**
EBITDA ($mil)	907.25	431.00	3442.40	984.10	2,535.15**	553.10**
EBIT ($mil)	526.71	204.97	2,557.91	466.84	2,031.20**	261.87**
Total Debt ($mil)	3,003.62	880.26	4,561.88	2,944.00	1,558.26	2,063.74***
Total Assets ($mil)	8,206.73	3,414.04	19,639.33	8,951.00	11,432.60**	5,536.96***
Cash to Total Assets	0.034	0.01	0.03	0.02	-0.01	0.01
Debt to Total Assets	0.37	0.37	0.35	0.33	-0.02	-0.05
Return on Assets	2.40	5.02	7.66	4.96	5.26*	-0.07*
Market to Book	2.05	1.85	2.03	1.86	-0.02*	0.01*
Free Cash Flow ($mil)	-117.47	-22.23	249.91	-125.52	367.38	-103.297

Table 4.2 reports the summary statistics of synergy-disclosing and non-disclosing bidder firms. All items are retrieved from the Thomson Financial database and are measured at the end of the fiscal year preceding the M&A announcement. The mean (median) significance is tested with the two-sided *t*-test (Wilcoxon rank-sum test). ***, **, and * indicate statistical significance at the 1%, 5%, and 10% levels, respectively.

In addition, the study requires a group of control firms which were not engaged in M&A during the respective time fame. A common benchmark index such as the S&P 500 subsector Energy is avoided, as this index does not control for M&A events, which is likely to bias the analysis of these events. More importantly, the analysis of its composition reveals that 36% of the energy firms in the present sample are member firms of this particular benchmark index, which will systematically bias the findings. Therefore, an index was manually created to serve as the benchmark index to identify abnormal capital market, as well as operating performance, following the merger. To be identified as a group of peer firms, they had to:

I. be based in the US
II. have primary SIC codes of either 1311, 4911, 4922-25, 4931-32, or 4941
III. not have been involved in an M&A transaction during the period of 1998 and 2008
IV. have available financial statements from 1995 to 2011
V. obtain a minimum sales volume of $500m and minimum total assets of $500m.

Criterion V ensures the exclusion of small firms, which experience varying capital market attention or diverging operational performance. The retrieved firms are then weighted on a daily basis by their market value to create a non-M&A, value-weighted, industry and size-adjusted index.

4.3.2 Valuation of Synergy Disclosures

As there is no binding obligation to disclose synergies, there is also no coherent method of doing so; thus, disclosure information varies with respect to the initial time of the synergy realization, as well as the time horizon. To ensure the comparability of these disclosures, the following assumptions were considered, and the communicated synergies were discounted to $t=0$; i.e. the announcement of the M&A. For those transactions which do not provide a detailed synergy disclosure, and instead solely forecast total synergetic gains, the following assumptions were made:

I. the synergies reach their steady state amount in year four, after completion (consistent with the majority of transactions that disclose a time horizon);
II. they grow thereto at a rate of 100%; and

III. the midpoint is considered if a synergy value range is indicated.

The synergies are discounted by following a procedure similar to Kaplan and Ruback (1995), Houston et al. (2001), Dutordoir et al. (2010), as well as Bernile and Bauguess (2010). Annual synergies, *S*, are assumed to realize in perpetuity at their steady state in year *T* while growing at the long-term inflation rate *I* (i.e. the 10-year inflation forecasted by the Federal Reserve Bank of Philadelphia at the time of the announcement), and to be subjected to the riskiness of the bidding firm's equity. The bidding firm's cost of equity is therefore computed with the capital asset pricing model, as the discount rate *K*. For this, the beta is estimated by regressing the bidder daily returns against the non-merging, industry-adjusted benchmark index in the time window of -20 to -273; in line with prior studies, the risk premium is assumed to be 7.5% with a corporate tax rate of 36%; as the risk-free rate the 10-year US government bond yield is used.

$$PV(SYNERGIES) = \sum_{t=i}^{T} \frac{S_t \times (1-0.36)}{(1+K)^t} + \frac{S_{i+T} \times (1-0.36)}{(K-I) \times (1+K)^{i+T}} \tag{4.1}$$

i denotes the days between the announcement and completion, divided by 365. Present value synergy gains are then scaled with the combined bidder and target market capitalization to control for the firm size. Appendixes A and B show a practical example on how to utilize the above-described model and its assumptions.

4.3.3 Econometric Design

Subsequent firm efficiency research and its results highly depend on the chosen research method. To capture varying value implications, multiple research measures are employed. With respect to the capital market's M&A valuation, the market model event study methodology is utilized, with the event in $t=0$ being the merger announcement (see Fama et al., 1969; MacKinlay, 1997). Cumulative average abnormal returns are calculated as the difference between the actual and expected returns of each firm for different time windows. Expected returns are estimated for the period of a full trading year from -20 to -273 days. To further exclude a systematic market bias which methodically underestimates takeover returns, two event windows are introduced, following Powell and Stark (2005), which start at five days prior to the event to five and ten days following the merger. The heteroskedasticity consisted Boehmer-Musumeci-Poulsen-test is used to test the results for statistical significance (Boehmer et al., 1991) as well as the non-parametric Wilcoxon rank-rum test.

To detect abnormal operating performance, four different performance measures and two instruments are utilized – i.e. the intercept and change model. The performance indicators consist of the operating cash flow (OCF) to book value of assets measure, following Healy et al. (1992). OCF are defined as sales (SDC item 01001) minus cost of goods sold (SDC item 01051) plus depreciation, amortization, and depletion (SDC item 01151), and changes in net working capital (SDC item 04900). The resulting OCF is scaled by the book value of total assets (SDC item 02999). Scaling the OCF by the market value of assets instead would bias the ratio downwards, according to Gosh (2001). In contrast, Betzer and Goergen (2009) argue that accounting measures (i.e. the book value of assets) are likely to be systematically biased, as acquiring firms may opt to use certain accounting policies and thereby affect the actual value. Therefore, scaling OCF by the book value of assets results in ratios which eventually do not show the true performance. To overcome this discrepancy, OCF is additionally scaled by the bidder market value of assets (SDC item 18100) as the second performance measure, in order to show potential varied implications. As the third performance indicator, the bidder return on assets (ROA) (SDC item 08326) is employed; this is a frequently utilized performance measure (e.g. Maiga & Jacobs, 2008), and is also employed by Becker-Blease et al. (2008) for the energy sector. This parameter hence makes it possible to directly assess the findings. Nonetheless, Krishnan, Krishnan, and Lefanowicz (2009) argue that sales and, respectively, return on sales (ROS) is a superior performance indicator. The benefit of sales it that it is a current measure (such as assets), yet is less sensitive to unexpected economic or industry factors, and is not subject to any accounting policies, in contrast to assets. Therefore, the final performance indicator employed is the bidder ROS. By utilizing these multiple performance indicators, it may also be possible to shed light upon potential measuring biases.

In the initial and primary analysis, the universe of non-merging energy firms is employed as a peer group to detect abnormal operating performance of merging firms. This is in line with Barber and Lyon (1996), who suggest that test statistics are best specified if firms with similar pre-merger performance are matched. Thus, in the first analysis each M&A-pursuing firm is matched with the entire size-matched peer group; in the second, they are matched with the respective performance quartile of the peer group, which is measured in the years prior to the merger, following Loughran and Ritter (1997). Therefore, the average, annual performance ratios of the peer firms (or respectively peer quartile) are initially computed, and this is then subtracted from the performance ratio of each merging firm i ($RATIO_i^{adj}$). To conduct a diversified analysis, the abnormal performance is measured for numerous pre- and post-merger periods, while the median values of the time periods are utilized instead of means to avoid the effects of extreme observations, which is in line with Gosh (2001). Abnormal operating performance is then allocated by regressing each post-merger, industry-adjusted performance measure by its pre-merger counterpart.

$$RATIO^{adj}_{post,i} = \alpha + \beta \times RATIO^{adj}_{pre,i} + \varepsilon_i \tag{4.2}$$

The expected increase in performance is indicated by the intercept α. Following Healy et al. (1992), this model is argued to be superior to the change model (the second research model) as it accounts for a possible persistence in cash flows measured by β. Therefore, the analysis will focus on this model.

As the second performance analysis, raw pre-merger performance ratios are assessed against post-merger ones for each of the four measures, and for different time periods prior to and following the merger. For each pre-merger period, mean and median returns for firm *i* are computed and assessed against their counterparts of firm *i*. Statistical significance for the mean and medians are inferred using the standard *t*-test and Wilcoxon rank-sum test, respectively.

4.4 Results and Discussion

4.4.1 Synergy Forecasts in the US Energy Sector

Overall, 45% of the energy firms in the sample disclose quantitative synergy estimates; this is fairly consistent with prior research by Bernile and Bauguess (2010), which finds a disclosure rate of 31% for the US energy sector, yet for a different time frame (i.e. 1990 to 2005); and the prior chapter, which also retrieves a disclosure rate of 45% for the EU energy sector. To further assess the magnitude as well as synergy drivers, these estimates are initially discounted to the time of the announcement, and then scaled by the combined bidder and target market capitalization 20 days prior to the announcement to exclude any valuation effect of the market due to the merger. Thereby, the average (median) beta of the firms in the sample is 0.98 (0.94), while the discount rate sums up to 12.56% (11.91%). From this it can be inferred that the energy industry risk is fairly consistent with the market risk, i.e. a beta of 1. Pure synergy figures finally amount to an average (median) of \$277.7 million (\$58.99 million). These figures are smaller than those found by general industry studies conducted by Bernile and Bauguess (2010), which find present values of \$828.7 million (\$189.7 million), yet larger than those of the real-estate sector, which are \$116.8 million (\$65.9 million) (Anderson, Medla, Rottke, & Schiereck, 2013).

While exploring synergy value drivers, the following OLS regression model is employed using the heteroskedasticity consistent test procedure by White (1980)

$$PV(SYNERGIES) = \alpha_0 + \alpha_1 BIDDER\_SIZE_i + \alpha_2 REL\_SIZE_i + \alpha_3 VALUE_i + \alpha_4 GEO_i + \alpha_5 DIVERSIFICATION_i + \alpha_6 YEAR_i + \alpha_7 SALES\_GROWTH\_B_i + \alpha_8 SALES\_GROWTH\_T_i + \alpha_9 LT\_DEBT\_B_i + \alpha_{10} LT\_DEBT\_T +_i \varepsilon_i \quad (4.3)$$

where the variables can be defined as in Table 4.3.

Table 4.3: Definition of variables

Variable	Definition
PV(SYNERGIES)	Present value of the disclosed synergies at the announcement of the M&A scaled by the combined market capitalization.
BIDDER_SIZE	Natural logarithm of the bidder assets in the year of the M&A announcement.
RELSIZE	Relative size as the ratio of target-to-bidder sales.
VALUE	Value of the M&A transaction in $ billion.
GEO	Geographic overlap allocates variable from 0 to 2, where 2 is attributed to mergers within the same state, 1 to those within the same region (Mid-Atlantic, Mid-West, New England, Central, Southeast, South-West, and West), and 0 otherwise.
DIVERSIFICATION	Business diversification is a dummy variable allocating 1 to deals with the identical SIC classification, and 0 otherwise.
YEAR	Year of the M&A announcement.
SALES_GRWOTH_B	Bidder sales of the year of the M&A divided by the sales four years ago.
SALES_GRWOTH_T	Target sales of the year of the M&A divided by the sales four years ago.
LT_DEBT_B	Bidder's long-term debt percentage of total capital.
LT_DEBT_T	Target's long-term debt percentage of total capital.

In addition, the utilized variables are checked for normality (as the sample size is 25) and multicollinearity. To confirm that the residuals of the regressions are normal distributed the Jarque-Bera test procedure is computed (Jarque & Bera, 1980). The evidence suggestst that the residuals are normal distributed. As a check for multicollinearity the VIF for each parameter are calculated and indicated in Table 4.4 below. Typically, for a VIF in excess of 10, serious multicollinearity is evident (Marquardt, 1970; Hair et al., 1995). As the largest VIF for the utilized data is 3.96, there is no indication of multicollinearity.

Table 4.4: Variance inflation factors

Variables	VIF
BIDDER_SIZE	3.958
RE_SIZE	1.884
VALUE	3.804
GEO	1.685
DIVERSIFICATION	2.580
YEAR	1.704
SALES_GROWTH_B	1.230
SALES_GROWTH_T	1.157
LT_DEBT_B	2.408
LT_DEBT_T	1.498

Table 4.5 below describes the above-defined regression model. Model 1 follows the outlined perpetuity assumption in computing the present values of the synergy forecasts, which is also in line with prior literature. Model 2 relaxes this assumption and expects synergies to realize five years after M&A completion. Thereby, an estimate of the effects of the perpetuity assumption is given, and the results are checked for robustness.

The evidence from Table 4.5 suggests that *REL_SIZE, SALES_GROWTH_B* as well as *LT_DEBT_T* have a significant influence on the amount of expected synergies within the energy sector. This correlation remains consistent when changing the discount model assumptions. These findings are interpreted via the explanation that bidders who have not yet exhausted their growth potential are able to extract a higher amount of synergies from M&A. Thus, energy mangers believe the M&A to enable an appreciation of their business, as a negative correlation is found between the sales growth and the amount of synergies, i.e. if the sales grew by a lesser extent, more synergies are expected. Whether the deal will actually increase sales and operating performance is part of the following investigation. Not surprisingly, the long-term debt of the target firm is negatively correlated with the amount of expected synergies. Hence, the acquirer expects that targets with lesser debt lead to more efficiency gains, and thus the integration and value potential is larger.

Table 4.5: Synergy value drivers

	(1) Perpetuity Model	(2) 5-Year Model
Intercept	-17.967	9.071
	(-1.005)	(1.415)
BIDDER_SIZE	-0.045	-0.010
	(-1.034)	(-0.343)
REL_SIZE	0.342**	0.234**
	(2.554)	(2.694)
VALUE	-0.001	0.000
	(-0.404)	(-0.120)
GEO	-0.006	-0.001
	(-0.368)	(-0.097)
DIVERSIFICATION	-0.020	-0.001
	(-0.460)	(-0.040)
YEAR	0.009	-0.005
	(1.002)	(-1.427)
SALES_GROWTH_B	-0.082**	-0.051**
	(-2.345)	(-2.512)
SALES_GROWTH_T	-0.067	0.008
	(-1.176)	(0.311)
LT_DEBT_B	0.010	-0.007
	(0.091)	(-0.098)
LT_DEBT_T	-0.154**	-0.103**
	(-1.907)	(-2.027)
Adjusted R^2	0.967	0.022
N	25	25

Table 4.5 shows the OLS regression on the discounted and scaled synergy forecasts of synergy-disclosing US energy firms. In Model 1, synergies are expected to grow in perpetuity and at the long-term inflation rate. Model 2 assumes synergies to realize five years after M&A completion. The variables are defined in Table 4.3. All computed *t*-statistics are robust with regards to heteroskedasticity. ***, **, and * indicate statistical significance at the 1%, 5%, and 10% levels, respectively

The current evidence regarding the relative firm size effect during mergers is mixed with respect to general industry observations. Seth (1990) finds that size differences lead to higher synergy gains, as smaller buyers can increase their market power, as well as enable economies by a greater extent, while merging with larger targets. In contrast, Ahuja and Katila (2001) provide evidence that firms of similar size can more efficiently integrate the merging entities, and thus lead to greater synergetic gains. With respect to the US energy sector, the evidence suggests that similar firm size leads to higher synergy gains. The positive regression coefficient exhibits that an increased target-to-bidder ratio (which results in increased size relatedness) leads to higher expected synergies.

Interestingly, *GEO* and *DIVERSIFICATION* in the US energy sector do not influence the expected amount of synergies. From this it can be inferred that energy sector managers perceive the sub-sectors and respective characteristics in the energy industry to be alike, and therefore multi-utility M&A do not affect the synergy potential.

4.4.2 Abnormal Capital Market Performance

Consistent with prior cross-industrial M&A research (e.g. Morck et al., 1990; Moeller et al., 2005), but also specifically for the energy market (e.g. Leggio & Lien, 2000), the sample shows in Panel A of Table 4.6 that all bidders experience a loss as their abnormal returns are, on average, negative, while those of all target firms are positive. The value loss for acquiring firms ranges from -3.27% to -3.64%, depending on the respective event window. However, if only those deals which disclose synergies are considered, and thus those transactions that are expected to create value through their merger, this wealth loss significantly reduces to -2.60% and -1.70%. The regression analysis in Panel B, allocating a dummy variable to synergy disclosing firms, confirms this finding and shows a significantly positive coefficient for disclosing firms.

As shown in the preceding section, synergy-disclosing bidders are significantly larger than to non-disclosing ones; Therefore, firm size was controlled for as a check for robustness in an additional analysis which also includes the firm size variable (not shown). Thereby, the synergy-disclosing dummy variable remains significantly positive, which further supports the findings and shows that the results are not driven by firm size. It is therefore shown that general market findings by Dutordoir et al. (2010), along with Bernile and Bauguess (2010) are also valid within the US energy sector, and that synergy disclosures are an efficient instrument to decrease existing information asymmetries. In addition, the study provides further support for Hutton et al. (2003), who argue that positive announcements lead to increased stock returns if they are disclosed along with quantitative, forward-looking statements. Thus, these forecasts enable rational market participants to perceive the value of the transaction. Interestingly, the effects of the synergy disclosure do not pass through to target firms. Here, all target firms, as well as the respective subsamples, gain significantly positive announcement returns, yet no capital market valuation differences are retrieved, owing to the synergy disclosure. Therefore, the market apparently allocates all efficiency gains to bidding firms, which further underlines the defined objective to analyze the operational performance of these firms in particular; the capital market expects the bidding entity to experience an increase in operational performance in the post-merger horizon.

Table 4.6: Synergy disclosure influence

Panel A: Cumulative average abnormal returns				
	[-5;+5]		[-5;+10]	
	Mean	Median	Mean	Median
All bidders	-3.64%***	-3.27%***	-4.86%***	-4.31%***
Disclosing bidders	-1.70%*	-2.60%	-2.02%*	-2.58%
Non-disclosing bidders	-4.51%***	-4.43***	-6.15%***	-5.41%***
Difference	2.81%*	1.83%*	4.49%**	2.83%**
All targets	18.87%***	13.20%***	17.91%***	12.66%***
Disclosing targets	20.98%***	11.61%***	20.60%***	12.24%***
Non-disclosing targets	17.91%***	13.73***	16.68%***	12.91%***
Difference	3.07%	-2.12%	3.92%	-0.67%
Panel B: Regression analysis				
	Bidder CAR		Target CAR	
Variables	[-5;+5]	[-5;+10]	[-5;+5]	[-5;+10]
Intercept	-0.045***	-0.062***	0.179***	0.167***
SYNERGY DISCLOSURE DUMMY	0.028**	0.041**	0.031	0.039
R^2	0.046	0.051	0.001	0.002
N	80	80	80	80

Table 4.6, Panel A presents descriptive statistics for bidder and target CARs for the [-5;+5] and [-5;+10] event window estimated using the market model event study methodology. The disclosing samples comprise deals in which bidding managers disclose a synergy forecast, and non-disclosure deals make up the remaining transactions. The mean (median) significance is tested with the two-sided *t*-test (Wilcoxon rank-sum test). Panel B reports the OLS regression for bidder and target CAR. Synergy disclosure is a dummy variable attributing 1 to deals with an associated synergy estimate. All computed *t*-statistics are robust with regards to heteroskedasticity. ***, **, and * indicate statistical significance at the 1%, 5%, and 10% levels, respectively.

The clear optimistic valuation of synergy-disclosing firms in the energy sector is apparently based on the prospect that the merger will result in the continuing realization of synergies, which finally results in the appreciation of firm value. However, as Becker-Blease et al. (2008) show, M&A-pursuing energy firms in general underperform compared to their counterparts, i.e. non-M&A-pursuing energy firms, when it comes to operational performance in the post-merger horizon. However, their operational performance analysis does not differentiate between disclosing and non-disclosing transactions, and therefore ignores the M&A objective. Yet, as the analysis above shows, this factor can have a significance influence on the valuation. In addition, within their performance measuring methodology Becker-Blease et al. (2008) solely use the change model. A more differentiated analysis with additional performance indicators will eventually present a more diversified perspective on post-merger performance, in particular as the analysis distinguishes between synergy-disclosing and non-disclosing firms.

4.4.3 Abnormal Operating Performance of Merging and Non-Merging Firms

The following section investigates whether synergy-disclosing M&A transactions experience a performance increase relative to non-merging firms. Table 4.7 therefore initially utilizes the intercept model, as suggested by Healy et al. (1992), to identify abnormal operating performance, in contrast to non-merging energy firms; the results for synergy-disclosing firms and non-disclosing firms are described in Panels A and B, respectively. Here, raw performance measures are adjusted by a sized-matched, non-merging industry portfolio, as well as a performance-matched, non-merging portfolio. Thereby, the operating performance is controlled for general and economic industry variations.

Panel A in Table 4.8 shows a negative abnormal operating performance for synergy-disclosing transactions. These remain robust against changes in the performance measure, benchmark composition, or in the pre-merger and post-merger horizon. However, none of these figures are statistically different from zero; thus, conclusive evidence is provided that synergy-disclosing transactions do not experience a decrease in their operating performance in the post-merger horizon. In contrast, as Panel B shows, non-disclosing transactions experience negative abnormal returns, which yet are statistically significant for Models 1, 3, and 4. Utilizing deflated OCF by total assets as the performance benchmark, negative intercepts are retrieved, which range from -1.91% to -2.61% for the size-matched portfolio and increase over a longer post-merger horizon. The ROA and ROS performance indicators reveal a comparable indication with even higher performance losses up to -7.69% and -24.15%, respectively, in the immediate year preceding the M&A. Utilizing the performance-matched benchmark, this value depreciation further increases. However, here these losses reduce over a longer post-merger horizon, which indicates that non-disclosing firms are able to stabilize their operating performance once fully integrated. In addition, it should be acknowledged that the results with respect to the ROS cannot necessarily be redirected to a loss in performance. Sales may not directly measure the productivity of a firm's assets, as they can, for instance, increase (which indicates a performance improvement) if a firm reduces its price level, though the asset base remains similar (Powell & Stark, 2005). Nonetheless, this ratio provides another indication, which is in line with other performance indicators and therefore further supports the notion that only non-disclosing firms suffer from M&A.

When utilizing OCF deflated by market value of assets, the abnormal performance remains negative for certain periods (and even becomes marginally positive); however, no intercepts can be distinguished from zero. It is likely that the lack of statistical significance can be allocated to the underlying attributes of market values, as this value is a forward-looking measure (Barber & Lyon, 1996). Potential gains, or in this case losses, relating back to the merger are immediately priced in the market value at the announcement (assuming efficient markets). Hence, the resulting market value not only incorporates the current assets (such as the BV of assets), but also all gains or losses to the assets the firm is expected to make in the future.

Overall, these findings suggest that synergy-disclosing firms, with the objective of increasing firm efficiency, do not experience negative operating performance in their post-merger horizon relative to other, non-merging energy firms. This is in contrast to the results retrieved by Becker-Blease et al. (2008), who conclude that M&A in the energy sector lead to value loss, i.e. negative abnormal operating performance. The results clearly show this not to be the case for those particular transactions which, *ex ante*, predefine the objective to increase efficiency and, moreover, pro-actively disclose this to their shareholders. Hence, merging energy firms can structure transactions such that no performance loss manifests, and eventually lead to the realization of synergy gains. For the subsample which does not disclose synergies at the M&A announcement, the results are in line with prior research and show a value loss in the post-merger period relative to the industry portfolio. Thus, these transactions apparently follow varied merger objectives. Consequently, the increased short-term abnormal capital market valuation for synergy-expecting transactions appears to be justified to the degree that no shareholder loss manifests.

Table 4.7: Abnormal size- and performance-adjusted operating performance

		(1)		(2)		(3)		(4)	
		OCF/BV of Assets		OCF/MV of Assets		ROS		ROA	
		Size-matched	Perf.-matched	Size-matched	Perf.-matched	Size-matched	Perf.-matched	Size-matched	Perf.-matched
Panel A: Disclosing firms' industry-adjusted abnormal performance (%)									
Period	[-1] to [+1]	-2.40	-1.91	-0.27	-0.00	-3.62	-2.79	-0.19	0.89
(in years)	[-1] to [+1;+2]	-2.36	-1.78	-1.20	-1.00	-1.77	-1.03	-0.20	0.13
	[-1] to [+1;+3]	-2.13	-1.89	-0.74	-1.14	-2.95	-1.13	-0.97	-0.62
	[-2;-1] to [+1;+2]	-1.61	-1.16	-1.00	-1.19	-2.45	-1.52	-0.12	-0.13
	[-2;-1] to [+1;+3]	-1.57	-1.42	-0.58	-1.26	-3.54	-2.07*	-0.69	-0.66
	[-3;-1] to [+1;+3]	-1.14	-0.73	-0.42	-1.14	-3.24	-2.08*	-0.74	-0.66
Panel B: Non-disclosing firms' industry-adjusted abnormal performance (%)									
Period	[-1] to [+1]	-1.71	-2.84*	0.29	1.88	-24.15*	-26.81**	-7.69**	-6.87***
(in years)	[-1] to [+1;+2]	-1.91*	-1.42	2.47	3.37*	-13.77**	-15.58***	-3.98**	-3.44***
	[-1] to [+1;+3]	-2.51**	-1.65*	-0.24	0.66	-7.85	-9.53*	-2.44*	-1.83**
	[-2;-1] to [+1;+2]	-1.78*	-1.27	2.61	3.38*	-15.45**	-11.21**	-3.93***	-3.28**
	[-2;-1] to [+1;+3]	-2.54**	-1.71*	-0.26	0.76	-8.06	-4.05	-1.08***	-1.43**
	[-3;-1] to [+1;+3]	-2.61**	-1.24	0.19	1.01	-2.89	-4.29	-1.53**	-1.30**

Table 4.7 shows the size- and performance-adjusted operating performance ratios of M&A-conducting US energy firms for various time periods (in years) prior to and following the merger. Panel A considers the subsample which discloses a voluntary synergy forecast at the announcement, while Panel B incorporates the remainder. All computed t-statistics are robust with regards to heteroskedasticity. ***, **, and * indicate statistical significance at the 1, 5%, and 10% levels, respectively.

To check the results for robustness, total sales (SDC item 01001) is used as a deflator for OCF, as suggested by Gosh (2001). The results and subsequent implications are unchanged in a qualitative sense. Furthermore the results are checked using mean values instead of medians, and the findings remain qualitatively the same.

Utilizing the change model as the second research methodology, the findings and conclusions above can be concluded. Table 4.8 assesses raw, unadjusted performance measures of merging firms, while separately analyzing synergy-disclosing and non-disclosing transactions, as well as mean and median values. Here, each firm's benchmark to assess performance variations in the post-merger period is its pre-merger performance.

As Panel A shows, there is no statistically significant performance decrease following M&A for synergy-disclosing firms, regardless of the applied performance ratio, deflator, or time period. With respect to non-disclosing transactions (Panel B), a similar picture manifests as there is hardly any statistically significant increase or decrease in the performance while utilizing OCF as the performance measure. However, considering the ROA and ROS as the performance estimator, the evidence suggests that the non-disclosing subsample experiences a significantly inferior performance in the initial two years following the merger. Therefore, this respective performance measure apparently drives the findings with respect to the change model.

Besides, the results retrieved from the change model are found to be lower than those from the intercept model, though it is in line with prior research by Linn and Switzer (2001) and Powell and Stark (2005). Overall, further evidence is obtained that M&A in the energy sector is not value-destroying, as Becker-Blease et al. (2008) argue, while no significant decrease is found relative to non-merging firms regarding any performance indicator for synergy-expecting M&A in the post-merger periods.

Table 4.8: Raw operating performance

		(1) OCF/BV of Assets		(2) OCF/MV of Assets		(3) ROS		(4) ROA	
		Mean	Median	Mean	Median	Mean	Median	Mean	Median
Panel A: Disclosing firms' raw operating performance (%)									
Year	-3	28.21	28.66	27.40	30.80	13.53	16.48	4.06	4.62
	-2	26.90	23.30	26.38	26.45	4.79	14.41	1.53	5.63
	-1	26.95	26.28	25.55	25.18	17.07	17.01	6.22	5.48
	+1	26.11	22.70	27.04	27.73	16.42	14.72	6.26	4.97
	+2	27.10	25.86	28.33	28.39	20.60	13.90	6.69	4.48
	+3	26.91	26.67	32.34	31.97	13.83	15.40	3.90	3.88
Period	[-1] to [+1]	-0.84	-3.58	1.50	2.55	-0.66	-2.29	0.04	-0.51
(in years)	[-2;-1] to [+1;+2]	-0.32	0.67	1.72	0.85	7.58	0.74	2.59	0.29
	[-3;-1] to [+1;+3]	-0.64	0.69	1.32	0.68	4.02	-0.01	1.23	-0.83
Panel B: Non-disclosing firms' raw operating performance (%)									
Year	-3	26.45	26.88	26.22	23.10	7.34	18.19	3.48	5.39
	-2	29.04	28.10	26.43	27.57	13.59	17.84	4.46	5.14
	-1	29.85	26.10	26.17	23.38	26.35	24.77	7.45	5.97
	+1	26.43	26.00	28.74	25.97	-12.72	14.73	-2.63	4.44
	+2	28.48	27.92	34.23	29.75	19.39	20.67	5.25	5.45
	+3	24.91	23.10	27.09	25.50	14.86	19.10	4.38	4.52
Period	[-1] to [+1]	-3.42	-0.10	2.57	2.60	-38.97***	-10.04***	-10.08***	-1.53***
(in years)	[-2;-1] to [+1;+2]	-1.99	-2.57	5.18**	2.59	-16.59**	-4.87	-4.64***	-1.04**
	[-3;-1] to [+1;+3]	-1.55	-1.42	2.63	1.93	-3.47	-0.11	-1.54	-0.46

Table 4.8 shows the raw performance ratios of M&A-conducting US energy firms for multiple time periods (in years) prior to and following the merger. Panel A considers the subsample, which discloses a voluntary synergy forecast at the announcement, while Panel B incorporates the remainder. Mean (median) performance variations during two periods are tested for statistical significance with a standard *t*-test (Wilcoxon rank-sum test). ***, **, and * indicate statistical significance at the 1%, 5%, and 10% levels, respectively.

4.4.4 Abnormal Operating Performance of Disclosing and Non-Disclosing Firms

A subsequent analysis estimates the performance increase of synergy-disclosing firms relative to non-disclosing ones, instead of the non-merging market as in the preceding subchapter. Thus, the performance-adjusted median returns are utilized, as these have been proven to more accurately estimate value variations for various time periods following the merger. For the OLS regression, each performance measure is regressed against the respective average present values of the synergy forecasts. For those firms without synergy disclosures a value of zero is allocated, which allows us to capture the influence of the disclosure on the operating performance. Control variables are also included, though these are not show as they are solely used to control for their influence on the synergy disclosure. Thus, the underlying regression is described as follows, while the variables are also checked for multicollinearity, with the largest VIF being 4.975; thus, there is no indication of multicollinearity.

$$PERFORMANCE_i = \alpha_0 + \alpha_1 PV(SYNERGIES) + \alpha_2 BIDDER\_SIZE_i + \alpha_3 REL\_SIZE_i + \alpha_4 VALUE_i + \alpha_5 GEO_i + \alpha_6 DIVERSIFICATION_i + \alpha_7 YEAR_i + \varepsilon_i \quad (4.4)$$

The significant positive coefficient for *PV(SYNERGIES)* in Table 4.9 on the following page provides evidence that synergy-forecasting deals exhibit a significantly higher performance relative to non-disclosing ones. Thus, non-disclosing transactions are shown to underperform compared not only to the non-merging market, but also with respect to forecasting M&A transactions. This provides further evidence that energy firms can apparently structure their M&A in such a manner that they realize synergetic gains, and thereby outperform those transactions which do not structure their merger to do so. In the following sub-chapter, these performance-driving factors will be investigated. The insignificance of Model 2 utilizing OCF deflated by MV of assets is not surprising, given the above-identified insignificant effect relative to non-merging firms.

Considering the various time periods, the evidence suggests that the excess performance of synergy-disclosing M&A diminishes over a longer post-merger horizon. From this it can be inferred that non-disclosing firms are able to increase their performance by gaining experience, while at the same time synergy-disclosing firms may find it difficult to continuously generate disclosed synergies. Nevertheless, on the whole Table 4.9 provides further evidence that the excess capital market gains due to synergy disclosures are not based on market overreaction, but rather on operational reasonability.

Table 4.9: Synergy disclosure influence on operating performance

	(1) OCF/BV of Assets	(2) OCF/MV of Assets	(3) ROA	(4) ROS
Panel A: Performance over period [+1;+1]				
Intercept	-0.774***	-0.514***	-1.072**	-7.515**
	(-5.572)	(-3.129)	(-2.245)	(-2.376)
PV(SYNERGIES)	6.157*	-1.728	5.204*	17.197
	(1.908)	(-0.434)	(1.794)	(1.553)
Control Vars.	Yes	Yes	Yes	Yes
N	80	80	80	80
Adjusted R^2	0.24	0.103	0.1395	0.352
Panel B: Performance over period [+1;+2]				
Intercept	-0.436***	-0.149	-0.467*	-3.462**
	(-4.055)	(-0.873)	(-1.882)	(-2.162)
PV(SYNERGIES)	10.979***	-1.348	8.792**	29.154**
	(3.132)	(-0.236)	(2.591)	(2.354)
Control Vars.	Yes	Yes	Yes	Yes
N	80	80	80	80
Adjusted R^2	0.291	0.027	0.088	0.284
Panel C: Performance over period [+1;+3]				
Intercept	-0.332***	0.087	-0.710***	4.201***
	(-3.105)	(-0.306)	(-3.358)	(-2.906)
PV(SYNERGIES)	7.951*	2.874	2.767	17.273*
	(1.743)	(0.514)	(1.349)	(1.668)
Control Vars.	Yes	Yes	Yes	Yes
N	74	74	74	74
Adjusted R^2	0.193	0.079	0.466	0.535

Table 4.9 shows the OLS regression on the median performance and industry-adjusted operating ratios for all US energy firms over respective time periods (in years) following the merger. *PV(SYNERGIES)* captures the average synergies in the corresponding years following the M&A, which are scaled by the bidder market value 20 days prior to the announcement. The control variable includes the following: *BIDDER_SIZE, REL_SIZE, VALUE, GEO, DIVERSIFICATION,* and *YEAR*; the variables are defined in Table 4.3. The computed *t*-statistics are robust with regards to heteroskedasticity. ***, **, and * indicate statistical significance at the 1%, 5%, and 10% levels, respectively.

4.4.5 Interaction of Operating Performance and Synergy Disclosures

A final analysis estimates the management's ability to foresee an increase or decrease in operating performance for the synergy-disclosing firm subsample using the OLS regression model from above (i.e. Model 4.4), again utilizing heteroskedasticity-consistent standard errors (White, 1980). The four regression models in Table 4.10 are also checked for the normal distribution of the residuals utilizing the Jarque-Bera test procedure; the evidence suggests that the results are normally distributed.

To confirm the accurate performance increase, each median abnormal performance-adjusted ratio (as these are expected to more accurately capture performance variations) for the first three years following the merger is regressed against the average scaled synergies for the same time interval. As the study is only interested in the bidder firm at this stage, present value synergies are solely scaled by the bidder's market capitalization. There is a deviation from the initial present value assumptions, as follows:

$$PV(SYNERGIES) = \frac{1}{3} \times \sum_{t=1}^{3} \frac{S_t \times (1-0.36)}{(1+K)^t} \tag{4.5}$$

As shown in prior sections, synergy-disclosing M&A outperform non-disclosing ones, and the addition of further control variables helps to explain those characteristics which lead to improved post-merger operating performance or affect performance losses. Thus, the firm size effect is particularly investigated, as larger firms are frequently attributed with higher profitability (Penman, 1991; Fama & French, 1995). Thus, the intention is to show that the increased operating performance relative to non-disclosing M&A is not solely driven by this parameter.

As Table 4.10 shows, there is a statistical significant and a positive correlation between the abnormal operating performance and *PV(SYNERGIES)*, which is, moreover, robust against changes in the performance ratio. This finding seems to suggest that relatively higher disclosed synergies also result in a more substantial increased operating performance. Therefore, energy managers are apparently able to estimate, *ex ante*, post-merger firm performance. The regression coefficient further implies, with respect, for instance, to OCF deflated by the book value of assets (Model 1), that an additional percentage point in scaled synergies leads to an 18% increase in the operating performance. While at first sight this appears to be relatively large, one has to consider that the scaled synergies for the first three years are, on average, just 0.17%. Therefore, firms are more likely to increase these by an additional 0.1%, which in turn would result in an operating performance increase of 1.8%. This ratio appears to be relatively stable across all performance indicators, except for ROA. Here, an additional 0.1% of scaled synergies leads to an improved operating performance of 0.6%. Form this it can be inferred that the realization of synergies will primarily increase the firm's sales, as well as OCF and, to a lesser extent, assets.

Table 4.10: Relation between synergy disclosure and operating performance

	(1) OCF/BV of Assets	(2) OCF/MV of Assets	(3) ROA	(4) ROS
Intercept	-2.536	15.74*	2.766	1.591
	(-0.230)	(2.005)	(0.723)	(0.307)
PV(SYNERGIES)	18.723**	30.846***	6.110**	16.192***
	(2.401)	(5.165)	(2.755)	(3.109)
BIDDER_SIZE	0.065	0.108*	0.039**	0.080*
	(0.836)	(1.823)	(2.588)	(1.816)
REL_SIZE	-0.185	-0.366**	-0.028	0.038
	(-0.763)	(-2.613)	(-0.656)	(0.491)
VALUE	-0.001*	-0.001	-0.001	-0.004*
	(-0.200)	(-0.277)	(-1.522)	(-1.909)
GEO	0.013	0.019	0.006	0.019**
	(0.487)	(1.027)	(1.158)	(2.334)
DIVERSIFICATION	0.106***	0.064**	0.031***	0.083***
	(2.914)	(2.250)	(3.748)	(3.697)
YEAR	0.001	-0.008*	-0.001	-0.001
	(0.205)	(-1.993)	(-0.762)	(-0.393)
Adjusted R^2	0.114	0.238	0.132	0.238
N	25	25	25	25

Table 4.10 shows the OLS regression on the performance and industry-adjusted operating ratios for synergy-disclosing US energy firms in the initial three years following the merger. *PV(SYNERGIES)* captures the average synergies in the initial three years following the M&A, which are scaled by the bidder market value 20 days prior to the announcement. The remaining variables are defined in Table 4.3. ***, **, and * indicate statistical significance at the 1%, 5%, and 10% levels, respectively.

The evidence concerning the control variables shows that *DIVERSIFICATION* in particular, i.e. whether the merging entities obtain identical SIC classification, matters for the improvement of operating performance. The significant positive coefficient implies that related lines of business lead to a significantly higher increase in the respective operating performance, regardless of the utilized performance ratio. This is in coherence with prior research by, for instance, Montgomery and Singh (1987). It is similarly argued here that due to related operational procedures and business processes, these firms have an increased potential to transfer knowledge, which subsequently leads to improved performance. In addition, unit costs and distribution channels can more easily be shared. Surprisingly, mangers in the energy sector perceive business relatedness differently, as shown in Table 4.5. Here, this parameter has no effect on the expected amount of synergies that they apparently have. Consequently, even though the management perceives the various energy sub-sectors to be similar, the evidence suggests that sector differences are a significant factor driving operational performance.

Interestingly, *REL_SIZE* does not influence the operational performance, except for in Model 2. Therefore, in contrast to general industry observations made by Seth (1990) and Ahuja and Katila (2001), size differences in the US energy sector do not influence operational performance in the post-merger horizon. The non-significance of the bidder and target size is consistent with prior research (Capon, Farly, & Hoenig, 1990; Robins & Wiersema, 1995; O'Leary-Kelly & Flores, 2002). This finding may, however, be partially driven by the fact that synergy-disclosing US energy firms (as shown in Table 4.5) are on average relatively large. Yet, again, energy-sector managers' perceptions regarding value drivers (see Table 4.5) diverge from the actual performance indicators, as they expect the relative size to significantly influence the to-be-realized synergies.

Table 4.11: Relation between synergy disclosure and operating performance (robustness)

	(1) OCF/BV of Assets	(2) OCF/MV of Assets	(3) ROA	(4) ROS
Panel A: Market premium to 6.5%				
Intercept	-2.509	15.614*	2.741	1.538
	(-0.227)	(1.978)	(0.718)	(0.296)
PV(SYNERGIES)	18.388**	30.000***	5.942**	17.769***
	(2.429)	(5.162)	(2.754)	(3.159)
Control Variables	Yes	Yes	Yes	Yes
Adjusted R-square	0.115	0.237	0.132	0.237
Panel B: Market premium to 8.5%				
Intercept	-2.571	15.878*	2.787	1.637
	(-0.234)	(2.031)	(0.727)	(0.316)
PV(SYNERGIES)	19.060**	31.695***	6.279**	16.618***
	(2.374)	(5.163)	(2.755)	(3.061)
Control Variables	Yes	Yes	Yes	Yes
Adjusted R-square	0.113	0.239	0.133	0.238
Panel C: Synergy present value synergies scaled by combined market capitalization				
Intercept	-7.129	8.537	1.839	-1.272
	(-0.704)	(1.084)	(0.546)	(-0.264)
PV(SYNERGIES)	22.227*	37.945***	9.343***	23.273**
	(1.728)	(3.698)	(3.073)	(2.790)
Control Variables	Yes	Yes	Yes	Yes
Adjusted R-square	0.092	0.168	0.151	0.242
N	25	25	25	25

Table 4.11 shows the robustness tests for Table 4.10. Panel A changes the risk premium assumption for the present value calculation of the synergy estimates to 6.5%, and Panel B changes the assumption to 8.5 %. Panel C scales the discounted synergy present values by the combined bidder and target market capitalization 20 days prior to the announcement. The control variables include the following: *BIDDER_SIZE, REL_SIZE, VALUE, GEO, DIVERSIFICATION,* and *YEAR*; the variables are defined in Table 4.3. The computed *t*-statistics are robust with regards to heteroskedasticity. ***, **, and * indicate statistical significance at the 1%, 5%, and 10% levels, respectively.

Furthermore, the results show that the *GEO* variable does not affect the post-merger operational performance, except with respect to ROS (Model 4). Here, a significantly positive coefficient is retrieved, which implies that the closer the two merging entities are located, the higher are the ROS. As this parameter is solely sensitive to sales, the finding suggests that energy firms can bundle their sales force operations more efficiency if they are organized and concentrated in one area. Most likely, energy firms can thereby exhibit market power to a certain degree, which subsequently leads to proportionally higher sales in comparison to an unrelated organization.

As a check for robustness, the initial assumptions with respect to present value synergy discloser calculations are revisited. The risk premium assumption is first deviated from 7.5% to 6.5% and 8.5%, which results in a changed discount rate K (see Panels A and B in Table 4.11, respectively). The results remain qualitatively unchanged. Moreover, disclosed synergies are scaled by the combined bidder and target market capitalization, instead of solely the bidder market capitalization, 20 days prior to the announcement (see Panel C); again the results and respective implications do not vary.

4.5 Implications

Based on the results, the following section develops recommendations for managers seeking M&A in general, as well as specifically for the US energy sector. In the initial analysis capital markets are shown to appreciate the disclosure of additional information; the evidence suggests that existing information asymmetries are thereby reduced, and thus the value perception increases. Therefore, managers should disclose further deal-related information with the intention to convey a value appreciation. Synergy forecasts are an efficient instrument by which to do this, and are especially suitable if they include quantitative forecasts. The resulting increased announcement returns are particularly important for mergers, as they increase the probability of deal completion (Luo, 2005). To generate the highest synergy gain from M&A, managers should look for targets of a comparable size. Similar organizational structures apparently enable a more instant and efficient integration. In addition, the results reveal that targets with fewer long-term debts enable an increased synergy potential.

Subsequent firm integration and operational performance diverges between firms which specifically structure their M&A to generate efficiencies (i.e. disclose synergies) and those that do not, while eventually following varied merger objectives. An essential driver of this increased operational performance is the business relatedness of the merging entities. Similar operational procedures not only enable instant improvement, but also excess performance. Therefore, managers should seek targets with comparable operations, best possible from the same industry. However, for certain industries, such as the US energy sector, even sub-sectors are dissimilar in the degree to which they hinder operational performance. Yet, if managers specifically aim to diversify their business while following an unrelated M&A strategy, geographic proximity has been shown here to also drive value. Thus, it is recommended that this M&A strategy specifically concentrates on the ability to consolidate sales operations. Sales are conditional on the market power within a region, and hence, if the target is able to increase this parameter through a similar regional focus, the ROS increase proportionally.

Finally, managers, particularly within the energy sector, should reconsider the parameters by which they estimate synergy gains from the merger. There are additional factors which drive value in the post-merger period, though these do not drive management-indicated synergy forecasts.

4.6 Conclusions

The results of this chapter highlight, in line with prior research, that capital markets attribute increased announcement returns to synergy-disclosing transactions. These transactions experience an increase of 2.8% relative to non-disclosing M&A. Thus, the voluntary synergy disclosure serves as an efficient tool to decrease existing information asymmetries between the market and the firm, and enables rational market participants to perceive the value of the M&A. The investigation of the value drivers of these synergy forecasts reveals that managers in the US energy sector perceive the relative firm size between the target and bidder to significantly influence the to-be-expected synergies; i.e. the more alike the firm size is, the higher the disclosed synergy gains. Moreover, the target's long-term debt has been proven to be a value driver, as less debt is associated with increased synergy potential.

The subsequent operating performance analysis which utilizes two methodologies (i.e. the change and intercept model) and four performance indicators shows that synergy-disclosing transactions do not experience a value loss relative to non-merging energy firms. In contrast, for the remaining non-disclosing M&A deals significantly worse performance ratios are found relative to the industry peer group, ranging from -2.51% for OCF deflated by the book value of assets to -15.6% for ROS for the initial three years following the M&A (utilizing the intercept model and performance-adjusted benchmark index). The study thereby shows, contrary to prior energy sector operating research, that M&A do not automatically lead to value loss. Merging firms can, in fact, prevent this value destruction if they structure their transaction and following integration respectively. In addition, the study provides evidence that synergy forecasting M&A generate significantly better operating performance relative to non-disclosing deals. Consequently, the increased short-term capital market valuation for synergy-disclosing transactions appears to be justified, and is therefore not based on management overconfidence or market overreaction.

Finally, the study investigated whether disclosed synergies correlate with operational performance in the initial years following the M&A, and, moreover, which deal characteristic synergy-disclosing M&A exhibit. Thereby, it is possible to identify those aspects which prevent value losses, as well as estimate whether managers are able to foresee, *ex ante*, the performance increase. The evidence presented here suggests that managers are indeed able to do so. In particular, an additional 0.1% in scaled synergies, for instance, leads to a 1.8% increase in the OCF deflated by the book value of assets. Moreover, the business relatedness has been proven to significantly increase the operating performance; thus, the more alike the operational procedures are, the higher the respective post-merger performance ratio. The geographic proximity matters only for sales-related performance ratios, from which it can be inferred that sales experience a particularly abnormal increase if the merger is concentrated in one area.

Interestingly, the results also reveal that the characteristics managers believe to drive value (i.e. increase synergy forecasts) may not necessarily be the driving force for operational performance. While management projections are highly conditional on the relative firm size, this parameter does not influence the operational post-merger performance. On the contrary, the significant influence of the business diversification on the post-merger operational performance does not have any significance for expected synergy estimates.

5 Concluding Remarks

This dissertation analyzes the impact of voluntary synergy disclosures at the M&A announcement on M&A transactions in the European, as well as US, energy sector. Thereby, not only is the capital market perspective considered on both a long- and short-term horizon, but the operational point of view is also taken into account. The relevance of this research mainly builds on two considerations: synergies are the primary driver of M&A, yet their role at the merger announcement has so far hardly been explored. Moreover, if managers decide to disclose synergy forecasts, the question of their operational reasonability arises; hence, whether these projections are actually realized, or are fostered by management overconfidence.

In responding to these general research issues, the dissertation addresses three points. Initially, Chapter 2 summarizes the literature on synergies studies in the energy sector. Thus, it presents the current academic evidence which is then extended according to the practical perspective of energy managers. Chapter 3 captures the capital market's valuation of energy-related M&A in Europe, and thereby also measures the effects of a voluntary synergy disclosure. Chapter 4 considers a varied perspective, and focuses on internal firm effects of synergy-expecting, energy-related M&A in the US. Thereby, the operational feasibility of synergy disclosures is evaluated.

The first empirical study (Chapter 2) of this dissertation focuses on the German energy market. The focus is laid on Germany as it is not only the largest electricity market in the EU, but also has not been researched thoroughly in the past, while the majority of synergy-related research has been completed for the US economy. Hence, synergy studies are advanced towards a novel marketplace, while also introducing a new methodical component, i.e. expert interviews. These contain open-ended questions and were conducted with the three leading, publicly traded energy providers in Germany. Combined, these managers have 17 years of M&A experience in the energy sector. The key benefit of this approach is that it does not report *ex post*, realized synergies (as empirical studies do) but rather *ex ante*, expected ones. This procedure hence makes it possible to analyze thus-far unexplored aspects.

Overall, the evidence suggests that financial synergies are of minor importance. Scale and scope economies are especially relevant for smaller utility firms, while these synergies emerge particularily in distribution and generation functions. Operational synergies, which are perceived to be the second most relevant synergy source, also emerge in distribution functions, yet also in the retail segment.

The second study (Chapter 3) investigates whether voluntary synergy disclosures are an efficient instrument by which to decrease information asymmetries. If so, increased short-term announcement returns would be seen, relative to those firms not disclosing synergy forecasts. To detect abnormal stock returns, the short-term event study methodology is utilized, while M&A announcements are hand collected and each firm's expected amount of synergies are individually discount to the time of the announcement. Therefore, 60 M&A are analyzed for the period 1998 to 2010, with 45% of the transaction-disclosing synergy forecasts. Additionally, the study tests whether potential increased short-term returns remain robust over a long-term period, utilizing the FF3F model, as well as the BHAR methodology. Hence, overall, this dissertation extends the literature on voluntary disclosure generally, and synergies at the merger announcement specifically.

The results show that transactions that are expected to lead to synergy gains and disclose these projections proactively to the capital market gain significantly higher announcement returns. Such a positive effect can, however, only be found for the bidding entity: acquirers experience increased short-term returns while disclosing synergy forecasts; targets, on the other hand, do not. Nevertheless, the combined entity does lead to value creation, and hence synergy disclosures efficiently reduce information asymmetries. Furthermore, in the long-term these firms are shown to be able to sustain their abnormal performance. Thus, the market does not overreact in the short-term, and management projections are apparently provided in a reasonable manner.

The third study (Chapter 4) extends the prior research to the US energy market. Initially, the capital market's response to voluntary synergy disclosures is again investigated in order to confirm the effects found for the EU in another environment for 80 transactions during the period 1997 to 2008. The subsequent objective of this chapter is to explore the operational reasonability of these synergy forecasts within the initial three years following the M&A. Therefore, an operationally driven analysis is conducted, excluding the market's valuation while utilizing two methodologies, i.e. the intercept and change model. Thereby, M&A are distinguised by their motive to realize efficiency gains, and it is expected that only these transactions will experience a performance increase. In addition, the characteristics that management believes drive the realization of synergies are investigated, along with those factors which actually foster a performance increase.

The evidence confirms that voluntary synergy disclosures in the US energy sector also lead to increased returns at the announcment of the M&A. The operating performance study shows that the M&A motive distinctively drives the post-merger performance. Those transactions which do not expect synergies not only underperform against size- and performance-matched non-merging firms, but also those firms that did disclose synergies. These, in contrast, show no significant difference relative to comparable firms; hence, no performance loss manifests. Thus, M&A can lead to value generation if the transaction and following integration is structured respectively. In addition, the results show that what management believes to drive value, such as for instance the relative target and bidder size, is not necessarily the case. This parameter reveals a strong influence on the management synergy projections, yet no effect on the operational post-merger performance.

To sum up, this thesis provides a comprehensive assessment of short- and long-term success of energy-related M&A in the European and US market from two different perspectives. It determines that capital markets attribute increased returns to those M&A transactions which expect synergy gains. As the post-merger accounting study confirms the realization of these synergy gains, it may be inferred that markets do not overreact and management boards are not over-confident. Therefore, if the merger allows for synergy gains, the communication strategy should be aligned in order to extract the maximum value perception from M&A.

Appendix

Appendix A: Synergy Disclosure Example – Devon Energy Corp.

The following provides an example of a management-released synergy disclosure at the merger announcement. In this case, a pure annual synergy figure is indicated, without a timeframe. In the following, the respective usage of the applied present value methodology is demonstrated.

Devon Energy Corp. announced a merger with Mitchell Energy & Development on 14 August 2001: *"[...] it (the M&A) expects to see annual operating and overhead synergies of at least $20 m."*

As there is no further information on the time horizon, it is assumed that the synergies will fully realize four years after completion, i.e. in 2007, as the merger was completed on 24 January 2002. Besides, the synergies are expected to grow until 2007 at a rate of 100%. In year four, the amount of synergies is assumed to remain perpetual (as there is no other indication in the management forecast). Hence, the following cash flow scheme evolves:

Table A. 1: Synergy realization horizon – Devon Energy Corp.

T=0 (2002)	T=1(2003)	T=3 (2004)	T=3 (2005)	T=4 (2006)	T=5 (2007)	...
1.25	2.5	5	10	20	20	...

All figures are in million USD.

The projected synergy gains are then discounted to retrieve their present value at the announcement. As 163 days passed between the announcement and the merger completion on 24 January 2002 (i.e. 0.45 years) the year-end realized synergies in the initial year, 2002, are discounted by a factor of 1.45. The discount rate of Devon Energy Corp. during this period equals 5.62%, while the inflation estimate during this period is 2.5%; thus, the following present value calculation emerges as:

$$PV(SYNERGIES) = \frac{1.25\times(1-36\%)}{(1+5.62\%)^{1.45}} + \frac{2.5\times(0.64)}{(1.0562)^{2.45}} + \frac{5\times(0.64)}{(1.0562)^{3.45}} + \frac{10\times(0.64)}{(1.0562)^{4.45}} + \frac{20\times(0.64)}{(1.0562)^{5.45}} + \frac{\$20mio\times(1-0.36)}{(5.62\%-2.5\%)\times(1+5.62\%)^{5.45}} \quad (A.1)$$

Appendix B: Synergy Disclosure Example – Wisconsin Energy Corp.

The following provides another example of a management-released synergy disclosure at the merger announcement. In this case, a specific time frame is indicated with an approximate synergy volume. In the following, the respective usage of the applied present value methodology is demonstrated.

Wisconsin Energy Corp. announced a merger with WICOR Inc. on 28 June 1999: *"Combination-related savings are anticipated to be approximately $35 million annually beginning in 2001. Savings are projected from lower costs for fuel, materials and services through enhanced purchasing power, elimination of duplication through attrition and sharing of resources."*

As there is specific information on the time horizon, the synergies are assumed to fully realize in 2001. Up to that point, the synergies are expected to grow at a rate of 100%. In 2001, the amount of synergies is assumed to remain perpetual (as there is no other indication in the management forecast). Hence, the following cash flow scheme evolves:

Table A. 2: Synergy realization horizon – Wisconsin Energy Corp.

T=0 (2000)	**T=1(2001)**	**T=3 (2002)**	**...**
17.5	35	35	...

All figures are in million USD

The projected synergy gains are then discounted to retrieve their present value at the announcement. As 303 days passed between the announcement and the merger completion on 26 April 2000 (i.e. 0.83 years), the realized synergies in the initial year 2000 are discounted by a factor of 1.83. The discount rate of Wisconsin Energy Corp. during this period equals 19.98%, while the inflation estimate during this period is 2.5%; thus, the following present value calculation emerges as:

$$PV(SYNERGIES) = \frac{17.5 \times (1 - 36\%)}{(1 + 19.98\%)^{1.83}} + \frac{35 \times (0.64)}{(1.1998)^{2.83}} + \frac{35 \times (1 - 0.36)}{(19.98\% - 2.5\%) \times (1 + 19.98\%)^{2.83}} \qquad \text{(A.2)}$$

References

AGEB. (2012). *Auswertungstabellen zur Energiebilanz für die Bundesrepublik Deutschland 1990 bis 2011*. Arbeitsgemeinschaft Energiebilanzen e.V. Retrieved from http://www.ag-energiebilanzen.de

Ahuja, G., & Katila, R. (2001). Technological acquisition and the innovation performance of acquiring firms: a longitudinal study. *Strategic Management Journal, 22*, 197-220.

Anderson, R., Medla, H., Rottke, N., & Schiereck, D. (2013). Why do REIT mergers take place? Evidence from synergetic forecasts. *Forthcoming in Journal of Real Estate Finance and Economics*.

Andrade, G., Mitchell, M., & Stafford, E. (2001). New evidence and perspectives on mergers. *Journal of Economic Perspectives, 15*(2), 103-120.

Baginski, S., & Hassell, J. (1997). Determinants of management forecast precision. *Accounting Review, 72*(2), 303-312.

Baginski, S., Hassell, J., & Waymire, G. (1994). Some evidence of the news content of preliminary earnings estimate. *Accounting Review, 69*(1), 265-271.

Barber, B. M., & Lyon, J. D. (1996). Detecting abnormal operating performance: the empirical power and specification of test statistics. *Journal of Financial Economics, 41*, 359-399.

Becker-Blease, J. R., Goldberg, L. G., & Kaen, F. R. (2008). Mergers and acquisitions as a response to the deregulation of the electric power industry: value creation or value destruction? *Journal of Regulatory Economics, 33*, 21-53.

Bednarczyk, T., Schiereck, D., & Walter, H. (2010). Cross-border acquisitions and shareholder wealth: evidence from the energy industry in Central and Eastern Europe. *Journal for Eastern European Management Studies, 15*(2), 106-127.

Berkovitch, E., & Narayanan, M. (1993). Motives for takeovers: An empirical investigation. *Journal of Financial & Quantitative Analysis, 28*(3), 347-364.

Bernile, G., & Bauguess, S. (2010). *Do merger-related operating synergies exist?* Working paper. Retrieved from http://papers.ssrn.com/sol3/papers.cfm?abstract_id=642322

Berry, S. (2000). Excess returns in electric utility mergers during the transition to competition. *Journal of Regulatory Economics, 33*, 21-53.

Betzer, A., & Goergen, M. (2009). *Disentangling the link between stock and accounting performance in acquisitions*. Working paper. Retrieved from http://papers.ssrn.com/sol3/papers.cfm?abstract_id=1474614

BMI. (2011). Germany Infrastructure Report, Business Monitor International.

Boehme, R. D., & Sorescu, S. M. (2002). The long run performance following dividend initiations and resumptions: underreaction or product of chance? *Journal of Finance, 57*(2), 871-900.
Boehmer, E., Musumeci, J., & Poulsen, A. B. (1991). Event study methodology under event-induced variance. *Journal of Financial Economics, 30*, 253-272.
Bösecke, K. (2009). *Value Creation in Mergers, Acquisitions, and Alliances*. Wiesbaden: Gabler.
Bowen, R., Davis, A., & Matsumoto, D. (2002). Do conference calls affect analysts' forecasts. *Accounting Review, 77*(2), 285-316.
Brunner, R. (2004). Where M&A pays and where it strays: a survey of research. *Applied Corporate Finance, 16*(4), 63-76.
Bruton, G., Oviatt, B., & White, M. (1994). Performance of acquisitions of distressed firms. *Academy of Management Journal, 37*(4), 972-978.
Burns, P., & Weyman-Jones, T. G. (1996). Cost function and cost efficiency in elecricity distribution: a stochastic frontier approach. *Bulletin of Economic Research, 48*(1), 41-64.
Bushee, B. J., Matsumoto, D., & Miller, G. S. (2004). Managerial and investor responses to disclosure regulation: the case of reg FD and conference calls. *The Accounting Review, 41*(5), 617-643.
Campbell, R. (2002). Shareholder wealth effects in equity REIT restructuring transactions: Sell-offs, mergers, and joint ventures. *Journal of Real Estate Literature, 10*(2), 205-222.
Capon, N., Farly, J. U., & Hoenig, S. M. (1990). A meta-analysis of financial performance. *Management Science, 36*(10), 1143-1159.
Chaturvedi, K., Chataway, J., & Wield, D. (2007). Policy, markets and knowledge: Strategic synergies in Indian pharmaceutical firms. *Technology Analysis and Strategic Management, 19*(5), 565-588.
Christensen, L., & Greene, W. (1976). Economies of scale in U.S. electric power generation. *Journal of Political Economy, 84*(4), 655-676.
Chuk, E., Matsumoto, D., & Miller, G. S. (2012). *Assessing methods of identifying management forecasts: CIG vs. researcher collected.* Retrieved from http://papers.ssrn.com/sol3/papers.cfm?abstract_id=1446978
Considine, T. (2004). Climate change: impact on the demand for energy. *Encyclopedia of Energy, 1*, 393-401.
Corhay, A., & Tourani Rad, A. (1996). Conditional heteroskedasticity adjusted market model and an event study. *Quarterly Review of Economics and Finance, 36*(4), 529-538.
Croci, E. (2007). Corporate raiders, performance and governance in Europe. *European Financial Management, 13*(5), 949-978.

Cullmann, A., & Hirschhausen, v. C. (2008). From transition to competition. *Economics of Transition 16*(2), 335-357.

Daim, T. U., Schweinfort, W., Kayakutlu, G., & Third, N. (2010). Identification of energy policy priorities from existing energy portfolios using hierarchical decision model and goal programming. *International Journal of Energy Sector Management, 4*(1), 24-43.

Datamonitor. (2005). Market Watch Energy. *Datamonitor*, 149.

Diaz Diaz, B., Sanfilippo Azofra, S., & Lopez Gutierrez, C. (2009). Are M&A premiums too high? Analysis of a quadratic relationship between premiums an returns. *Quarterly Journal of Finance and Accounting, 48*(3), 5-21.

Dutordoir, R., Roosenboom, P., & Vasconcelos, M. (2010). *Synergy disclosures in mergers and acquisitions*. Working paper. Retrieved from http://papers.ssrn.com/sol3/papers.cfm?abstract_id=1571546

Dye, R. (1985). Disclosure of nonproprietary information. *Journal of Accounting Research, 23*(1), 123-145.

EC. (2007). *An Energy Policy for Europe*. European Commission, Brussels.

Edlin, A., & Stiglitz, J. (1995). Discouraging rivals: managerial rent-seeking and economic inefficiencies. *American Economic Review, 85*(5), 1301-1314.

Erickson, M., & Wang, S. (1999). Earnings management by acquiring firms in stock for stock mergers. *Journal of Accounting and Economics, 27*, 149-176.

Evans III, J. H., Hannan, R. L., Krishnan, R., & Moser, D. V. (2001). Honesty in managerial reporting. *Accounting Review, 76*(4), 537-559.

Fabrizio, K. R., Rose, N. L., & Wolfram, C. D. (2007). Do markets reduce costs? Assessing the impact of regulatory restructuring on US electric generation efficiency. *American Economic Review, 97*(4), 1250-1277.

Fama, E. (1998). Market efficiency, long-term returns, and behavioral finance. *Journal of Financial Economics, 49*, 283-306.

Fama, E., Fisher, L., Michael, J., & Roll, R. (1969). The adjustment of stock prices to new information. *International Economics Review, 10*, 1-21.

Fama, E., & French, K. (1993). Common risk factors in the returns on stocks and bonds. *Journal of Financial Economics, 33*, 3-56.

Fama, E., & French, K. (1995). Size and book-to-market factors in earnings and returns. *The Journal of Finance, 50*, 131-155.

Fama, E., & French, K. (1997). Industry cost of equity. *Journal of Financial Economics, 43*, 153-193.

Fee, E., & Thomas, S. (2004). Sources of gains in horizontal mergers: evidence from customers, suppliers, and rival firms. *Journal of Financial Economics, 74*(3), 423-460.

Ferris, S. P., & Park, K. (2001). How different is the long-run performance of mergers in the telecommunications industry? *Innovations and Investments in Corporate Finance, 7*, 122-144.

Filippini, M. (1996). Economies of scale and utilization in the Swiss electric power distribution industry. *Applied Economics, 28*(5), 543-551.

Fischer, P., & Verrchecchia, R. (2000). Reporting bias. *Accounting Review, 75*(2), 487-507.

Flamos, A., Grafakos, S., Qikonomou, V., & Zevgolis, D. (2010). Multi-criteria analysis weighting methodology to incorporate stakeholders' preferences in energy and climate policy interactions. *International Journal of Energy Sector Management, 4*(3), 434-461.

Francis, J., Schipper, K., & Vincent, L. (2002). Expanded disclosures and the increased usefulness of earnings announcements. *The Accounting Review, 77*(3), 515-546.

Fraquelli, G., Piacenza, M., & Vannoni, D. (2004). Scope and scale economies in multi-utilities: evidence from gas, water and electricity combinations. *Applied Economics, 36*, 2045-2057.

Frodel, M., & Schmidt, C. (2009). Am Tropf Russlands. *Perspektiven der Wirtschaftspolitik, 10*(1), 79-91.

Froelich, K., & McLagan, J. R. (2008). Diversification strategy in electric utilities: Who wins? Who looses? *Academy of Strategic Management, 7*, 1-20.

Froggatt, A., & Levi, M. A. (2009). Climate and energy security policies and measures: synergies and conflicts. *International Affairs, 85*(6), 1129-1141.

Geiger, S., & Hoffman, J. (1998). The impact of the regulatory environment and corporate level diversification on the firm performance. *Journal of Mangerial Issues, 10*(4), 439-453.

Gelb, D. S., & Zarowin, P. (2002). Corporate disclosure policy and the informativeness of stock prices. *Review of Accounting Studies, 7*, 33-52.

Gosh, A. (2001). Does operating performance really improve following corporate acquisitions? *Journal of Corporate Finance, 7*, 151-178.

Granier, L., & Podesta, M. (2010). Bundeling and mergers in energy markets. *Energy Economics, 32*, 1316-1324.

Gugler, K., Mueller, D., Yurtoglu, B., & Zulehner, C. (2003). The effects of mergers: an international comparison. *International Journal of Industrial Organization, 21*(5), 625-653.

Hair, J. F., Anderson, R. E., Tatham, R. L., & Black, W. C. (1995). *Multivariate Data Analysis* (Vol. 3). New York: Macmillan.

Hambrick, D., & Hayward, M. (1997). Explaining the premium paid for large acquisitions: Evidence of CEO hubris. *Administrative Science Quaterly, 42*(1), 103-127.

Harford, J. (2005). What drives merger waves? *Journal of Financial Economics, 77*(3), 529-560.

Hattori, T., Jamasb, T., & Pollitt, M. (2005). The electricity distribution in the UK and Japan: A comparative efficiency analysis. *Energy Journal, 26*(2), 23-47.

Healy, P., Hutton, A., & Palepu, K. (1999). Stock performance and intermediation changes surrounding sustained increases in disclosure. *Contemporary Accounting Research, 16*, 485-520.

Healy, P., & Palepu, K. (2001). Information asymmetry, corporate disclosure, and the capital markets: A review of the empirical literature. *Journal of Accounting and Economics, 31*, 405-440.

Healy, P., Palepu, K., & Ruback, R. (1992). Does corporate performance improve after mergers? *Journal of Financial Economics, 31*, 135-175.

Healy, P., Palepu, K., & Ruback, R. (1997). Which takeovers are profitable? Strategic or financial? *Solan Business Review, 38*, 45-57.

Hess, B. (2010). Evaluating the efficiency effects of industry consolidation - Evidence from US interstate pipeline companies. *International Journal of Energy Sector Management, 4*(3), 462-481.

Heston, S., Rouwenhorst, G., & Wessels, R. (1999). The role of beta and size in the cross-section of European stock returns. *European Financial Management, 5*(1), 9-27.

Houston, J., James, C., & Ryngaert, M. (2001). Where do merger gains come from? Bank mergers from the perspective of insiders and outsiders. *Journal of Financial Economics, 60*, 285-331.

Hutton, A., Miller, G., & Skinner, D. (2003). The role of supplementary statements with management earnings forecasts. *Journal of Accounting Research, 41*, 867-890.

IMAA. (2012). Global Energy Sector M&A, Institute of Mergers, Acquisitions and Alliances. From http://www.imaa-institute.org/

Jaggi, B. (1978). Comparative accuracy of management's annual earnings forecast. *Financial Management, 7*(4), 24-32.

Jandik, T., & Makhija, A. K. (2005). Can diversification create value? Evidence from the electric utility industry. *Financial Management, 34*(1), 61-93.

Jarque, C. M., & Bera, A. k. (1980). Efficient tests for normality, homoscedasticity and serial independence of regression residuals. *Economic Letters, 6*(3), 255-259.

Jensen, M. (1986). Agency costs of free cash flow, corporate finance, and takeovers. *American Economic Review, 76*(2), 323-329.

Jo, H., & Kim, Y. (2007). Disclosure frequency and earnings management. *Journal of Financial Economics, 84*(2), 561-590.

Kaplan, S., & Ruback, R. (1995). The valuation of cash flows: an empirical analysis. *Journal of Financial Economics, 50*, 1059-1093.

Kavanagh, R. (2002). Electric acid test – Germans plug into power privatization. *NEFTE Compass, 5*.

Kimbrought, M. (2005). The effect of conference calls on analyst and market underreaction to earnings announcements. *The Accounting Review, 80*(1), 189-219.

Kimbrought, M., & Louis, H. (2011). Voluntary disclosure to influence investor reactions to merger announcements: an examination of conference calls. *Accounting Review, 80*(1), 189-219.

Klimchuk, G. (1999). Mergers to continue on upswing. *Natural gas, 15*(8), 2-5.

Kode, G., Ford, J., & Sutherland, M. (2003). A conceptual model from evaluation of synergy in mergers and acquisitions: A critical review of literature. *South African Journal of Business Management, 34*(1), 27-38.

Krishnan, H., Hitt, M., & Park, D. (2007). Acquisition premiums, subsequent workforce reduction, and post-acquisition performance. *Journal of Management Studies, 44*, 709-732.

Krishnan, H., Krishnan, R., & Lefanowicz, C. (2009). Market perception of synergies in related acquisition. *Academy of Strategic Management Journal, 8*, 99-119.

Kwoka, J. (2005). Electric power distribution: economies of scale, mergers, and restructuring. *Applied Economics, 37*, 2373-2386.

Kwoka, J., & Pollitt, M. (2010). Do mergers improve efficiency? Evidence from restructuring the US electric power sector. *International Journal of Industrial Organization, 28*, 645-656.

Lang, M., & Lundholm, R. (1996). Corporate disclosure policy and analyst behavior. *Accounting Review, 71*, 467-492.

Lang, M., & Lundholm, R. (2000). Voluntary disclosure and equity offerings: Reducing information asymmetry or hyping the stock? *Contemporary Accounting Research, 17*(4), 623-662.

Larsson, R., & Finkelstein, S. (1999). Integrating strategic, organizational and human resource perspectives on mergers and acquisitions. *Organizational Science, 10*(1), 1-26.

Leggio, K., & Lien, D. (2000). Mergers in the electric utility industy in a deregulatory environment. *Journal of Regulatory Economics, 17*(2), 69-85.

Lev, B., & Penman, S. (1990). Voluntary forecast disclosures, nondisclosures, and stock prices. *Journal of Accounting Research, 28*, 49-76.

Linn, S., & Switzer, J. (2001). Are cash acquisitions associated with better post combination operating performance than stock acquisitions? *Journal of Banking and Finance, 25*, 1113-1138.

Lipin , S., & Sirowner, M. (2003). Investor communications: a message of confidence in the deal. *Mergers & Acquisitions, 38*(10), 49-76.

Loughran, T., & Ritter, J. R. (1997). The operating performance of firms conducting seasoned equity offerings. *Journal of Finance, 52*(5), 1823-2850.

Louis, H. (2004). Earnings management and the market performance of acquiring firms. *Journal of Financial Economics, 74*(1), 121-148.

Lubatkin, M., & Srinivasan, N. (1997). Merger strategies and stockholder value during times of relaxed antitrust enforcement: the case of large mergers during the 1980's. *Journal of Management, 23*(1), 59-81.

Luo, Y. (2005). Do insiders learn from outsiders? Evidence from mergers and acquisitions. *Journal of Finance, 60*(4), 1951-1982.

Lyon, J. D., Barber, B. M., & Tsai, C. (1999). Improved methods for long-run abnormal stock returns. *Journal of Finance, 54*(1), 165-201.

MacKinlay, C. A. (1997). Event studies in economics and finance. *Journal of Economic Literature, 35*(1), 13-39.

Madura, J., & Ngo, T. (2008). Clustered synergies in the takeover market. *Journal of Financial Research, 4*, 333-356.

Maiga, A. S., & Jacobs, F. A. (2008). Assessing JIT performance: an econometric approach. *Journal of Management Accounting Research, 20*, 47-59.

Marquardt, D. W. (1970). Generalized inverses, ridge regression, biased linear estimation, and nonlinear estimation. *Technometrics 12*, 591-612.

Matuszewski, L. J. (2010). Honesty in managerial reporting: Is it affected by perceptions of horizontal equity? *Journal of Management Accounting Research, 22*, 233-250.

Merton, R. C. (1987). A simple model of capital equilibrium with incomplete information. *Journal of Finance, 42*, 483-510.

Moeller, S., Schlingemann, F., & Stulz, R. (2004). Firm size and the gains from acquisitions. *Journal of Financial Economics, 73*(2), 201-228.

Moeller, S., Schlingemann, F., & Stulz, R. (2005). Wealth destruction on a massive scale? A study of acquiring-firm returns in the recent merger wave. *Journal of Finance, 60*, 757-782.

Moeller, S., Schlingemann, F., & Stulz, R. (2007). How do diversity of opinion and information asymetry affect acquirer return? *Review of Financial Studies, 20*, 2047-2078.

Montgomery, C., & Singh, H. (1987). Corporate acquisition strategies and economic performance. *Strategic Management Journal, 8*, 37-386.

Morck, R., Shleifer, A., & Vishny, R. (1990). Do managerial objectives drive bad acquisitions. *Journal of Finance, 45*, 31-48.

Mueller, D. (1969). A theory of conglomerate mergers. *Quaterly Journal of Economics 83*(4), 63-659.

Myers, S., & Majluf, N. (1984). Corporate financing and investment decisions when firms have information that investors do not have. *Journal of Financial Economics, 13*, 187-222.

O'Leary-Kelly, S. W., & Flores, B. E. (2002). The integration of manufacturing and marketing/sales decisions: Impact on organizational performance. *Journal of Operations Management, 20*, 221-240.

Officer, M., Poulsen, A., & Stegemoller, M. (2009). Target-firm information asymmetry and acquirer return. *Review of Finance, 13*, 467-493.

Painuly, P. P. (2009). Financing energy efficiency: lessons from experiences in India and China. *International Journal of Energy Sector Management, 3*(3), 293-307.

Penman, S. H. (1991). An evaluation of accounting rate of return. *Journal of Accounting, Auditing, and Finance, 6*(2), 233-255.

Pereira da Silva, P., & Soares, I. (2008). EU spot prices and industry structure: assessing electricity market integration. *International Journal of Energy Sector Management, 2*(3), 340-350.

Piacenza, M., & Vannoni, D. (2009). Vertical and horizontal economies in the electrical utility industry: an integrated approach. *Annals of Public and Cooperative Economics, 80*(3), 431-450.

Powell, R. G., & Stark, A. W. (2005). Does operating performance increase post-takeover for UK takeovers? A comparison of performance measures and benchmarks. *Journal of Corporate Finance, 11*, 293-317.

Ramos-Real, F. J., Tovar, B., Iootty, M., Fagundes de Almeida, E., & Queiroz Pinto Jr., H. (2009). The evolution and main determinants of productivity in Brazilian electricity distribution 1998–2005: An empirical analysis. *Energy Economics, 31*, 298-305.

Robins, J., & Wiersema, M. F. (1995). A resource-based approach to multibusiness firm: Empirical analysis of the portfolio interrelationships and corporate financial performance. *Strategic Management Journal, 16*, 277-299.

Roll, R. (1986). The hybrids hypothesis of corporate takeovers. *Journal of Business, 2*(1), 197-218.

Salvanes, K., & Tjotta, S. (1994). Productivity differences in multiple output industries: an empirical application to electricity distribution. *Journal of Productivity Analysis, 5*, 23-43.

Schwitzer, J. (1996). Evidence of real gains in corporate acquisitions. *Journal of Economics and Business, 48*, 443-460.

Seth, A. (1990). Value creation in acquisitions: a re-examination of performance issues. *Strategic Management Journal, 11*, 99-115.

Sing, M. (1987). Are combination gas and electric utilities multiproduct natural monopolies. *Review of Economics & Statistic, 69*(3), 392-399.

Stankeviciute, L., & Criqui, P. (2008). Energy and climate policies to 2020: the impact of the European "20/20/20" approach. *International Journal of Energy Sector Management, 2*(2), 252-273.

Studness, C. (1996). Utility diversification: Dismal past, uncertain future. *Public Utilities Fortnight, 134*(20), 44-45.

Sueyoshi, T., & Goto, M. (2011). Operational synergy in the US electric utility industry under an influence of deregulation policy: A linkage to financial performance and corporate value. *Energy Policy, 39*, 699-713.

Sueyoshi, T., Goto, M., & Shang, J. (2009). Core business concentration vs. corporate diversification in the US electric utility industry: Synergy and deregulation effects. *Energy Policy, 37*, 4583-4594.

Ting-Kun, L. (2010). An empirical study of firms' merger motivations and synergy from Taiwanese banking industry. *International Research Journal of Finance and Economics, 38*, 13-27.

Verde, S. (2008). Everybody merges with somebody - the wave of M&As in the energy industry and EU merger policy. *Energy Policy, 36*, 1125-1133.

White, H. (1980). A heteroskedasticity-consistent covariance matrix estimator and a direct test for heteroskedasticity. *Econometrica, 48*, 817-838.

Yatchew, A. (2000). Scale economics in electricity distribution: a semiparametric analysis. *Journal of Applied Econometrics, 15*(2), 187-210.

Zhang, H. (1998). US evidence on bank takeover motives: a note. *Journal of Business Finance & Accounting, 27*(7), 1025-1034.

Corporate Finance and Governance

Herausgegeben von Dirk Schiereck

Band 1 Sebastian Michael Gläsner: Return Patterns of German Open-End Real Estate Funds. An Empirical Explanation of Smooth Fund Returns. 2010.

Band 2 Patrick Ams: Directors' Dealings and Insider Trading in Germany. An Empirical Analysis. 2010.

Band 3 Joachim Vogt: Value Creation within the Construction Industry. A Study of Strategic Takeovers. 2011.

Band 4 Fabian Braemisch: Underpricing, Long-Run Performance, and Valuation of Initial Public Offerings. 2011.

Band 5 Matthäus Markus Sielecki: Creating and Governing an Integrated Market for Retail Banking Services in Europe. A Conceptual-Empirical Study of the Role of Regulation in Promoting a Single Euro Payments Area. 2011.

Band 6 Arne Wilkes: Determinants of Credit Spreads. An Empirical Analysis for the European Corporate Bond Market. 2011.

Band 7 Dirk Schiereck / Martin Setzer (Hrsg.): Bankerfolg und Akquisitionen. Aktuelle Erkenntnisse zur Konsolidierung und Restrukturierung der Finanzindustrie. 2011.

Band 8 Steffen Meinshausen: M&A Activity, Divestitures and Initial Public Offerings in the Fashion Industry. 2012.

Band 9 Malte Helmut Raudszus: Financial Return Risk and the Effect on Shareholder Wealth. 2012.

Band 10 Ramit Mehta: Mergers and Acquisitions in the Global Brewing Industry. A Capital Market Perspective. 2012.

Band 11 Marcel Normann: The Influence of German Top Executives on Corporate Policy and Firm Performance. 2012.

Band 12 Christoph Böhm: Risk-Adjusted Performance and Bank Governance Structures. 2013.

Band 13 Christian Thamm: Minority Shareholder Monitoring and German Corporate Governance. Empirical Evidence and Value Effects. 2013.

Band 14 Robert M. Fraunhoffer: Mergers and Acquisitions in the Energy Sector. The Impact of Synergy Disclosures on Shareholder Wealth and Operating Performance. 2013.

www.peterlang.de

www.ingramcontent.com/pod-product-compliance
Ingram Content Group UK Ltd.
Pitfield, Milton Keynes, MK11 3LW, UK
UKHW020837290926
15470UKWH00004B/1395

9 783631 645215